INTIMATE STRANGERS

UNSEEN LIFE ON EARTH

INTIMATE STRANGERS
UNSEEN LIFE ON EARTH

Cynthia Needham • Mahlon Hoagland • Kenneth McPherson • Bert Dodson

ASM PRESS
Washington, D.C.

ASM Press
American Society for Microbiology
1752 N Street, NW
Washington, DC 20036-2804

Library of Congress Cataloging-in-Publication Data

Intimate strangers : unseen life on earth / Cynthia Needham . . . [et al.].
p. cm.
"Based on the PBS television series Intimate strangers : unseen life on earth."
ISBN 1-55581-163-9
1. Microbiology--Popular works. 2. Microorganisms--Popular works.
I. Needham, Cynthia, 1946–

QR56.I56 2000
579--dc21

99-059802

Acknowledgments

The authors wish to thank colleagues and friends who contributed their time and ideas to this effort.

The television series *Intimate Strangers: Unseen Life on Earth* launched our work. It was planned and developed over several years by Peter Baker, Cynthia Needham, and many other contributing scientists and became the centerpiece for the Microbial Literacy Collaborative's initiative. The series was produced by Baker & Simon Associates under Peter's leadership as Executive Producer. We are grateful to Series Producer Julio Moliné; to Producers Carl Byker ("Tree Of Life"), David Mrazek and Peter Baker ("Keepers of the Biosphere"), Marlo C. Bendau and co-producer M. Scott Martin ("Dangerous Friends and Friendly Enemies"), and Mark Ritts and Cynthia Crompton ("Creators of the Future"); and to Production and Finance Manager Casey Spira for all his assistance with image research and rights.

We are also grateful to the funding organizations who made this initiative possible: The National Science Foundation, The American Society for Microbiology, The United States Department of Energy, The Annenberg/CPB Project, The Corporation for Public Broadcasting, The Arthur Vining Davis Foundation, and The Foundation for Microbiology.

Michael Goldberg and Susan Kee of the American Society for Microbiology (ASM) provided encouragement along with critical advice in the early development of each section. Ellie Tupper, our editor at ASM Press, was an anchor throughout the ebbs and flows of the creative process. Ellie's thoughtful advice, along with that of Jeff Holtmeier, Director of ASM Press, always allowed us to improve what we had written. Sue Ricker gave us valuable secretarial assistance throughout the project.

The following scientists shared their knowledge and expertise with us as the book progressed: Joan Bennett, Julian Davies, Stanley Falkow, William R. Green, Stephen Morse, Norman Pace, Javier Penalosa, Elmer Pfefferkorn, Elio Schaechter, Catherine Squires, James Tiedje, Carl Woese, and Ralph Wolfe.

This book is as much about images as it is about words. Ed Reinhardt of Blue Mango Productions oversaw the book's design and served as its primary photo illustrator. Mardi McGregor created the striking photo montages throughout the book. Molly Bowman, Marina Pacilio, Kim Sonderland, and Michael Balabayev provided assistance with the many original illustrations. Neil Blume and John Booth at Oregon Public Broadcasting coordinated the difficult job of transferring images from the video series to the book. Finally, Hank Woolsey and Precision Graphics pulled it all together.

Contents

SECTION ONE. KEEPERS OF THE BIOSPHERE 1

SECTION TWO. THE TREE OF LIFE 45

SECTION THREE. DANGEROUS FRIENDS AND FRIENDLY ENEMIES 89

SECTION FOUR. CREATORS OF THE FUTURE . . . 137

The World That Came in the Mail

The world arrived in the mail. It was marked "Fragile." I lifted it out of its package and cautiously placed it on the accompanying stand. I peered in.

I could see life in there—a network of branches, some encrusted with green filamentous algae, and six or eight small animals, mostly pink, cavorting, so it seemed, among the branches. In addition, there were hundreds of other kinds of beings, as plentiful in these waters as fish in the oceans of Earth; but they were all microbes, much too small for me to see with the naked eye.

Clearly the pink animals were shrimp. They caught your attention immediately because they were so busy. A few that had alighted on branches were walking on 10 legs and waving dots of other appendages. One was devoting all its attention, and a considerable number of limbs, to dining on a filament of green.

If you're in charge of a little world like this, and you conscientiously concern yourself about its temperature and light levels, then whatever you may have had in mind at the beginning—eventually you care about who's in there. If they're sick or dying, though, you can't do much to save them.

The ghostly molting shrouds and the rare dead body of an expired shrimp do not linger long. They are eaten, partly by the other shrimp, partly by invisible microorganisms that teem through this world's ocean. And so you are reminded that these creatures don't work by themselves. They need one another. The shrimp take oxygen from the water and exhale carbon dioxide. The algae take carbon dioxide from the water and exhale oxygen. They breathe each other's waste gases. Their solid wastes cycle also, among plants and animals and microorganisms. In this small Eden, the inhabitants have an extremely intimate relationship.

The shrimp's existence is much more tenuous and precarious than that of the other beings. The algae can live without the shrimp far longer than the shrimp can live without the algae. The shrimp eat the algae (and microorganisms), but the algae mainly eat light.

Unlike an aquarium, this little world is a closed ecological system. Light gets in, but nothing else—no food, no water, no nutrients. Everything must be recycled. Just like the Earth.

Our big world is very like this little one, and we are very like the shrimp. But there is at least one major difference: we are able to change our environment. We can do to ourselves what a careless owner of such a crystal sphere can do to the shrimp. If we are not careful, we can heat our planet through the atmospheric greenhouse effect or cool and darken it in the aftermath of a nuclear war. With acid rain, ozone depletion, chemical pollution, radioactivity, the razing of the tropical forests and a dozen other assaults on the environment, we are pushing and pulling our little world in poorly understood directions.

If we are not graced with an instinctive knowledge of how to make our technologized world a safe and balanced ecosystem, we must figure out how to do it. We need more scientific research and more technological restraint. It is probably too much to hope that some Great Ecosystem Keeper in the sky will reach down and put right our environmental abuses. It is up to us.

It should not be impossibly difficult. Birds—whose intelligence we tend to malign—know not to foul the nest. Shrimps with brains the size of lint know it. Algae know it. One-celled microorganisms know it. It is time for us to know it too.

Carl Sagan
Condensed from "The World That Came in the Mail" by Carl Sagan

Introduction to a Small, Small World

Imagine that, like Alice, you stepped through the Looking Glass and became small—VERY small, in fact, 1/200,000th of your normal size. The world around you would look very strange indeed, like some planet elsewhere in the universe, a land filled with odd inhabitants and mysterious forms and shapes.

There are lavender feather boas, a foot or more long, that twist and undulate past you, clearly intent on reaching some unknown destination. Swirls of green and purple pillow-sized blimps propel about, slowing to nudge you, then engaging their long propeller-like appendages to dart out of view. A glistening creature slides up, its changing shape seeming to envelop its surroundings in transparent ooze. It stops as though to look you in the eye and say good day, then moves on, leaving no trace behind.

Nearby a ring of pulsating bean bags, golden in color, move in an ever closing and then opening circle, like children on a playground. As you watch their ballet, their bodies combine to form a giant stalk that sprouts magnificent podlike structures. Across the way, you note the remnants of what must surely have been some devastating epidemic. The empty shells of the beings rock aimlessly, their silence a reminder that death must follow life here just as it does in your more familiar world.

The inhabitants of this strange and tiny world are microbes—bacteria, fungi, viruses, protozoa, and algae. Most people think of them as germs, but these intimate strangers are much more than that. They are the force that keeps our planet working.

The inhabitants of this normally invisible world, taken together, represent the largest mass of living creatures on earth. Their numbers likely exceed the number of stars in the universe. Although their individual actions have little impact, collectively they have changed the face of our planet, creating the oxygen-rich atmosphere essential for our survival. Their actions continue to drive the major chemical cycles that underpin life today. Their activities will shape all life in the future.

The inhabitants of this invisible world were here long before we were. In fact, their most ancient relatives were the first life forms on the planet. We have traced the whispers of their evolutionary history back over more than 3.5 billion years to a time when earth was a far different place than it is today. We can read their history in their genes, but our history is written in their genes as well, as is the history of all other living creatures. All of life's diversity evolved from their simple, single cells.

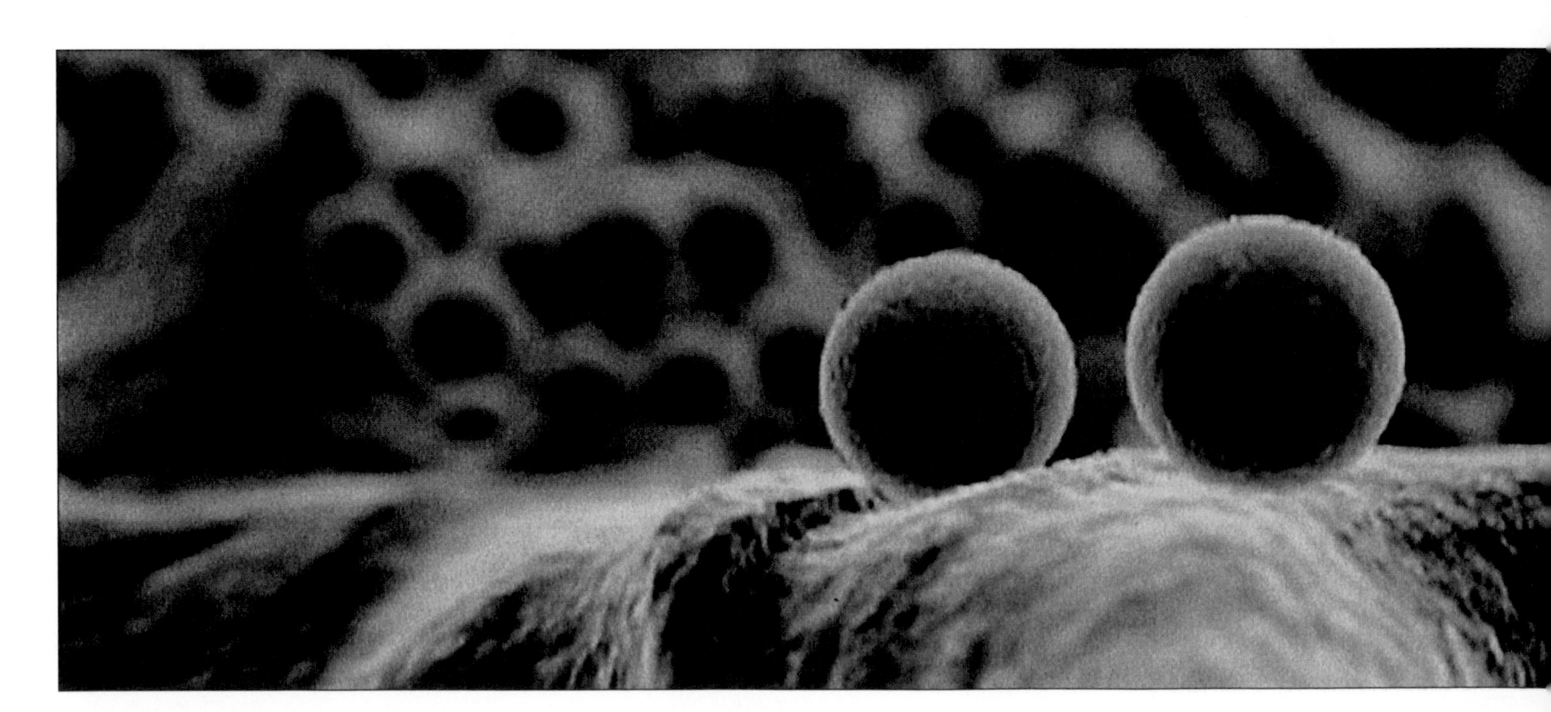

The inhabitants of this invisible world are not all entirely benign. Although most all of them are harmless companions, a few can suddenly turn against us, causing disease in us and in our plants and animals. Others are familiar enemies with which we have learned to live. And a very few are lethal strangers that can quickly overwhelm us unless we can mount a swift and powerful defense.

In spite of their almost 4-billion-year history on earth, our discovery of microbes is very recent. Knowledge of their existence came only after we had magnifying devices that would allow us to look into their unseen world—to step through the Looking Glass. In the mid-17th century, a Dutch amateur microscope builder, Antoni van Leeuwenhoek, was amazed and delighted when he first saw these "wee animalcules" through his carefully focused lenses. As microscopes improved, others were able to peer with increasing fascination at these tiny creatures.

But it was not until the very end of the 19th century that we came to know microbes as more than amusing oddities. It was then that we began to discover the extraordinary power of these tiny inhabitants and their role on our planet. Pasteur identified bacteria as the agents of spoilage—finally laying to rest the idea that life arose spontaneously from decaying matter. Koch and other scientists established that some of these tiny creatures were agents of disease. Their discoveries opened the door to pasteurization, sanitation, and immunization, practices that allowed us to prevent the microbes' unwanted and harmful activities. By the early 1950s, we had discovered antibiotics, microbial products that allowed us to cure previously fatal diseases.

Over the last half of the 20th century, our knowledge of this unseen world and its links with larger life has expanded exponentially. Scientists studying microbes learned that DNA was the genetic material, explored how the basic processes of life—such as reproduction and energy transformation—are carried out, and probed the evolutionary significance of these first of earth's inhabitants. Now scientists are racing to learn the secrets of their vast and interconnected communities—communities that make our life on earth possible—before we so disrupt their environment that they change the face of our planet again.

Many believe that our very future depends on our ability to understand this unseen world and how it works. With our increasing knowledge, we are entering into partnerships with the microbes, taking advantage of their skills to solve some of humanity's most difficult problems—treating and preventing infectious diseases, feeding an ever-growing population, and cleaning up our polluted environment.

This is the world of *Intimate Strangers: Unseen Life on Earth.* Enter their world with enthusiasm, and learn what you can from earth's oldest and most successful inhabitants.

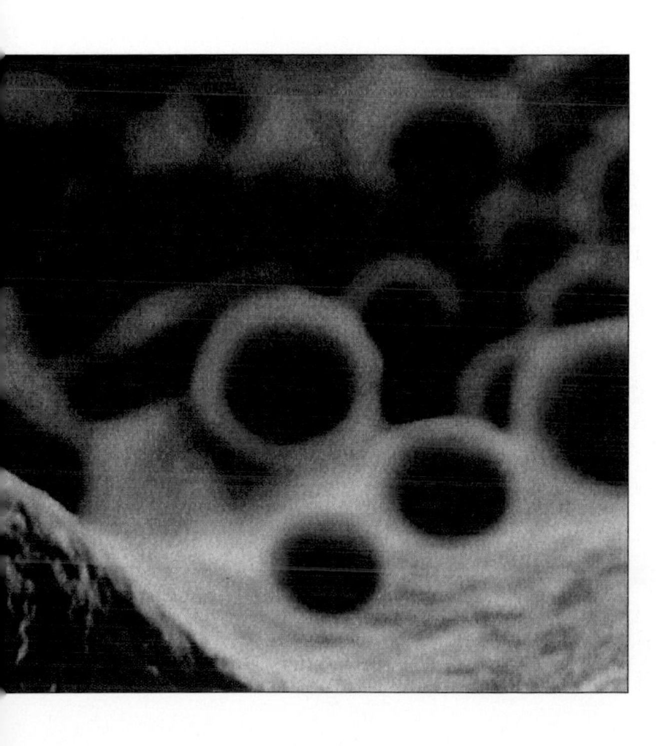

Denizens of the Microbial World

The microbial world includes all living organisms that can perform the basic functions of life—metabolism, reproduction, and adaptation—as single-celled creatures. We know these creatures as bacteria, archaea, and algae, fungi, and protozoans. This world also includes a striking anomaly—a string of nucleic acids, existing in its own Buckydome, that relies on other cells to handle the chores of everyday life. We know this ne'er-do-well as a virus.

The microbial world is a world of order and rules . . . and a world of striking contradictions. Microbes are so small they are invisible to the human eye, but one of the largest living organisms on earth—the huge Wisconsin underground mushroom—is an aggregate of microbes. Millions of microorganisms can fit into a teaspoon of seawater, but the microbial community makes up more than half of life's total mass. An individual microbe has virtually no impact on its community, but collectively, microbes have shaped the face of our planet.

You will meet some specific microbes in field guides to the dramatic players in our story. We have invested them with personalities to reflect their dominant role on the stage we call life. At this point in time—unless somebody has *again* redrawn the tree—scientists use common biological traits to assign these players to one of three main groups.

Bacteria and Archaea (a.k.a. the Prokarya)

The Bacteria and Archaea are the smallest independent living cells. They are on average 50 times smaller than our cells and 70 times larger than viruses.

These organisms all share a common structural feature—they lack a true nucleus. Their genome consists of a single molecule of double-stranded DNA, one end linked to the other to make a circle some 5 million nucleotides in length. The circle is twisted up and lies free in the cytoplasm. The total number of genes in an organism is a measure of its complexity, i.e., the number of proteins it needs to do its thing. The prokaryotes' 5 million nucleotides equal about 5,000 genes. We humans, by comparison, have some 80,000 genes.

A membrane similar to that in our cells surrounds the bacterial cell. Outside this membrane is a tough cell wall made of protein and carbohydrate. Many members of the Bacteria and Archaea have whiplike extensions (called flagella) that rotate, propelling them through their watery environment like an outboard motor. They are also able to form tubes (called pili) that can link up to other microbes or attach to other surfaces, such as the membrane of our epithelial cells. When they attach to like microbes, they use the pili to pass DNA from one to another (lateral gene transfer) in a single-celled version of sex. Bacteria and Archaea reproduce without sex, however, by dividing in half, a feat some can perform every 12 to 20 minutes under favorable conditions.

Bacteria generally reside in places familiar to us: for example, soil, water, food, and animal skin and digestive tracts. Some possess brightly colored pigments that allow them to use light as an energy source. Others have adapted to use virtually every other conceivable source of energy, including rock! Some require oxygen, just as we do, but many can live quite satisfactorily without it, and others avoid it as a toxic chemical.

The Archaea, in contrast, are often found in what we would consider the harshest of conditions—extremes of heat and salt, highly acid or highly alkaline environments—places where most other life forms would fear to tread. They avoid oxygen, and some of them engage in unusual biologic activities like producing methane (natural gas).

It could be said that these microbes—the Bacteria and Archaea—are the most successful forms of life, given their multi-billion-year durability and their occupancy of essentially every niche earth has to offer.

Algae, Fungi, and Protozoa (a.k.a. the Microbial Eukarya)

The microbial Eukarya are organisms that all contain a true nucleus. They are considerably larger, more complex, and more varied structurally and functionally than the Prokarya. Their DNA is spooled on proteins in structures called chromosomes and is harbored in their membrane-encased nucleus. Eukaryotic DNA contains some 10 times the number of genes found in the Prokarya. New evidence suggests that the Eukarya arose quite early in evolutionary history as well. The ancestors of the microbial Eukarya were most likely the progenitors of all plants and animals.

Eukarya mostly live as single cells, but some form colonies and some are truly multicellular. Kelp, for instance, is made up of permanent colonies of algae, and mushrooms are differentiated aggregates of fungus cells. Algae and fungi are surrounded by tough cell walls similar to the Bacteria and Archaea. Protozoa such as amoebae and paramecia are more animal-like than plant-like. Unlike algae and fungi, they lack rigid cell walls and are able to move about by various means and capture their food. Certain microbial Eukarya form the masses of microorganisms in the oceans, called plankton, that are a critical link in the food chain.

Viruses

Viruses are not cells. They are packages of DNA (or sometimes RNA) containing relatively few genes in a protein shell. They are roughly 2,000 times smaller than an animal cell. Sometimes a membrane picked up from the cells they infect surrounds them, and they can infect virtually all types of cells—bacterial, fungal, protozoan, plant, and animal. It seems likely that they arose relatively late in evolution, perhaps even from living cells.

Viruses can exist outside cells for long periods in a dormant state, but can only reproduce themselves inside living cells, subverting the cell's machinery for their own use. They may multiply to large numbers, destroying the host cell and causing disease. Others may enter and leave cells without causing trouble. And sometimes they leave a few of their genes behind. They may also incorporate a gene from their host into their own genomes and then deposit the gene in other cells they infect. Indeed, 1% of human DNA contains segments of viral DNA, suggesting such past associations.

SECTION ONE

KEEPERS OF THE BIOSPHERE

We call upon the earth, our planet home, with its beautiful depths and soaring heights, its vitality and abundance of life, and together we ask that it teach us and show us the way.

—Chinook Blessing

In the Arizona desert just north of Tucson a giant glass structure sits in the valley and looms over the landscape, catching the light from the sun. The Buckminster Fuller-like geometric arrays along its glass walls evoke a mental image of the pyramids of Egypt. The interior is enormous, eerie, silent and haunted. Once filled with teeming plant and animal life, the structure has become home to cockroaches and crazy ants. Much of the original plant life has died.

The structure's creators, funded by Texas billionaire Edward Bass, were drawn from the scientific and business communities. They envisioned their enterprise to be the first step toward human colonization of Mars. Their mini-world was to be self-sustaining, driven by the solar energy that penetrated its glass walls and by the power grid that supplied the electricity for heating, cooling, and running its 750 sensors. They designed it to contain a rainforest, ocean, plains, savanna, marsh, and desert. They named it Biosphere 2 to reflect its relationship to Biosphere 1—the earth.

In 1991, the greenhouse-like structure was completed. Eight individuals, later known as "Biospherians," sealed themselves inside to begin what was to have been a two-year attempt at self-sustained living. But much went wrong.

The surrounding mountains reduced the amount of sunlight available for growing Biosphere 2's plants. The double glass used to build it further reduced the amount of sunlight available, including the ultraviolet rays that are critical for the synthesis of certain vitamins and the reduction of certain plant parasites.

The visually stunning design complicated matters further by accumulating dust, trapping water, and fostering the buildup of bacteria and algae on the glass surfaces, all of which led to diminished solar power inside the structure. The designers used soils that encouraged excessive growth of microbes, resulting in higher than normal concentrations of carbon dioxide and nitrous oxide. The unsealed concrete walls absorbed free oxygen.

Over the course of the first several months after the Biospherians were sealed inside, the oxygen concentration dropped, and the carbon dioxide and nitrous oxide levels rose to dangerously high levels. Officials were forced to pump in oxygen and remove carbon dioxide in order to protect the lives of the individuals inside. Crop production was so poor that the occupants had food smuggled in to them. Both actions violated the tenet of isolation.

The critical balances among the chemical cycles that sustain us on Biosphere 1 were never achieved in Biosphere 2. The enterprise failed, in part, because Biosphere 2's creators didn't appreciate the complexity involved in constructing a closed, synthetic ecosystem where everything must recycle and remain in balance.

Where some saw a costly disaster, others saw opportunity, and today Columbia University has taken over Biosphere 2. Their scientists are redesigning its interior, making it into a working laboratory to study the very cycles of life that led to its initial downfall. The researchers are optimistic that a redesigned Biosphere 2 will allow them to answer questions about the large-scale cycles that are driven by earth's tiniest inhabitants, which dictate living conditions on earth. Their hope is to help us learn how to better manage our planet.

The remarkable balance of life on our planet is maintained through the activities of its tiniest inhabitants. The inhabitants are microbes—bacteria, fungi, protozoa, algae, and viruses. They keep materials cycling through the biosphere, making resources available for all living things. From the rainforests to the open oceans to the deep sea trenches, microbes are the glue that holds our web of life together.

2001, a Space Odyssey

It's early morning in Manhattan, one of the world's most beautiful urban areas. But it's not beautiful at the 59th Street Marine transfer station. Here a barge, piled high with open mounds of trash and circled by a halo of seagulls, makes its way out of the station and into the Hudson River. A small part of the city's more than 8,500 tons of daily garbage is on its way to the Fresh Kills Landfill on Staten Island. In the Bronx, workers load trash destined for a landfill in Amelia County, Virginia, the recipient of one of New York City's largest exports—its garbage. And so it goes in each of the other three boroughs that make up New York City.

New York City has a problem. Little of the enormous amounts of garbage its citizens create is recycled. It's merely buried in landfills. At the end of 2001, the Fresh Kills Landfill will close. The New York City Mayor offered a solution—ship it somewhere else. Unfortunately, his plan was not particularly appealing to the neighboring states. All of the potential recipients resisted the notion that New York City's substantial cultural achievements obligated its neighbors to accept its garbage. Nevertheless, 2001 looms near and New York City must find space to replace Fresh Kills.

While most of the materials associated with life are constantly moving through cycles of reuse, the by-products of "civilization" are accumulating in massive quantities.

Methanothrix

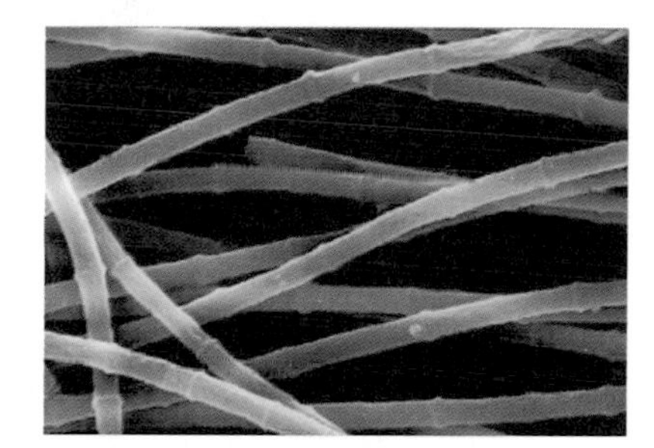

Identity: Bacterium
Residence: Sludge
Favorite pastime: Releasing methane
Activities: A member of The Methane Makers, *Methanothrix* operates in sewage to generate enough natural gas (methane) to heat a home.

New York City is not alone. We generate immense quantities of waste; most is not easily or quickly returned for use. Life in a large urban center magnifies the problem, but we all face a similar dilemma—finding ways to turn our garbage much more quickly back into useful materials.

Imagine what would happen to our planet if other living creatures followed our example, tying up biologic resources in forms that couldn't be readily used.

Fortunately for us, we are only one link in a much larger system on our planet, a system where everyone's garbage is someone else's meal. We can think of these linkages as a complex, multi-dimensional food web, an array of exchanges going on continuously among the communities of living creatures that populate the planet.

Life is constantly cycling all the resources necessary to grow and reproduce. And it's that cycling that sustains all of life on earth.

Life's Essential Ingredients
SUNLIGHT
HEAT
HYDROGEN
SULFUR
PHOSPHORUS
NITROGEN
OXYGEN
CARBON

Cycling Through the Biosphere

For all practical purposes, our beautiful blue-green planet is a closed system. Unlike Biosphere 2, we can't open a valve and pump in more oxygen, hydrogen, carbon, or nitrogen. The resources on earth are finite and scarce, and all its life must share—cycle—those resources to survive. Much of that cycling depends on earth's tiniest inhabitants, the microbes.

We tend to view each of earth's cycles as a separate, individual process—animals use oxygen and generate CO_2, plants use CO_2 and produce oxygen. In fact, all of earth's cycles interweave and interact to create an intricate, complex system that sustains the conditions necessary for life. Regulatory feedback mechanisms tie the cycles together. What changes one cycle—to a greater or lesser degree—changes all cycles.

Scientists are reasonably confident that they understand much about the movement of individual elements within individual cycles. They are less confident in their knowledge about the forces which regulate individual cycles. They are far less confident in their knowledge of the regulatory processes which govern the complex system of cycles that maintain the relative stability of living conditions on earth.

Consider the element carbon. Just like everything else living on earth, we are made of carbon atoms. Carbon is a part of our protein, our DNA, and our energy storage substances like sugar and fat. In fact, carbon accounts for almost a quarter of all the atoms in our body. Carbon, along with hydrogen, oxygen, and nitrogen, is a key ingredient in all organic (living) matter. Despite its prominence in living cells, most of the carbon atoms on our planet—estimated to be almost 20,000,000 billion tons of atoms—are in rock (inorganic matter). Most of this carbon is like the garbage in New York City's landfills—it doesn't cycle very rapidly. That means that living creatures must get the carbon necessary to make themselves from somewhere else.

It may come as a surprise that much of the carbon used by living creatures comes from the air. That carbon exists in the earth's atmosphere as carbon dioxide—a single atom of carbon attached to two atoms of oxygen.

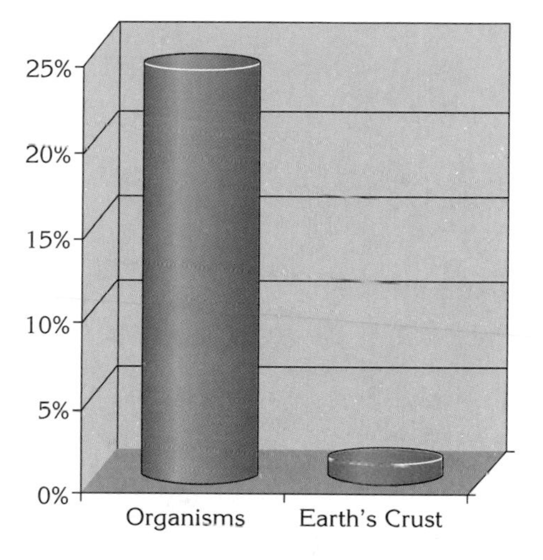

Life concentrates carbon
Although life contains only a tiny fraction of earth's total carbon, the amount of carbon relative to the total mass of atoms in living creatures is remarkably high.

Life's essential chemicals cycle constantly between living organisms and the land, water, and atmosphere in which they live. Energy enters from the sun and from earth's hot interior, then disperses, spent as heat, into outer space.

Chlorella

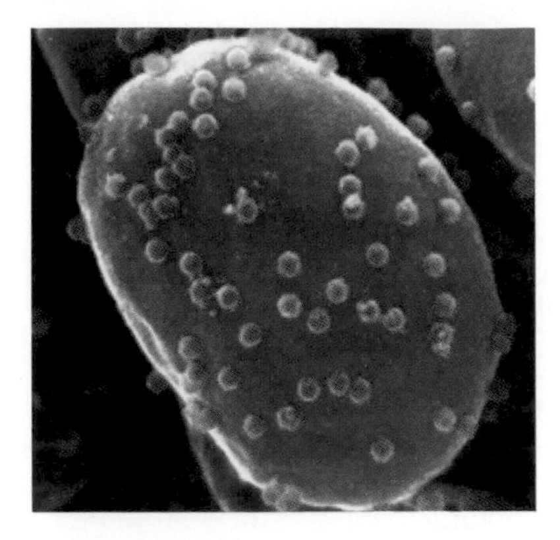

Identity: Alga
Residence: Ocean surface waters
Favorite pastime: Basking in the sun
Activities: A member of The Oxygen Producers, *Chlorella* floats on the ocean surface, photosynthesizing and helping to generate half the oxygen we breathe.

The Carbon Cycle

When life on earth was first emerging, living organisms found the carbon they needed to make their cell material in the water in which they lived. But this carbon was limited compared to the plentiful carbon dioxide in earth s atmosphere. Early in evolution, microbes, and later plants, gained the ability to convert carbon dioxide from the atmosphere into a biologically usable form in the process called photosynthesis. This was a giant leap forward.

Most people think only of plants when it comes to photosynthesis, but photosynthetic microbes like algae and cyanobacteria are just as important. Both plants and the photosynthetic microbes transform energy from sunlight into chemical energy, capturing carbon dioxide in the process. The plants and microbes use the sun s energy to combine the carbon dioxide with hydrogen from water to make glucose the sugar that ultimately provides energy and much of the building material necessary for living creatures.

A major fringe benefit of photosynthesis from our perspective is the release of the oxygen atoms from carbon dioxide back into the atmosphere. Oxygen, which has its own cycle, is the principal waste product of photosynthesis. Thus, plants and microbes not only remove carbon dioxide from the air, they also return oxygen.

The contribution of photosynthetic microbes may come as a surprise. They contribute as much as half of the total oxygen that we breathe.

Kelp, a kind of seaweed composed of communities of brown and red algae, makes up dense forests in the oceans. Algae and cyanobacteria use the sun's energy to make most of their substance from carbon dioxide.

Of course, converting carbon into sugar is only one stop on the carbon cycle, and microbes play many other roles in keeping this cycle running. In plants, much of the stored carbon goes to make cellulose, the material that gives plant cells their structure. Animals and insects—called grazers—that eat only plants have to be able to break down the cellulose in order to get enough energy and carbon to build their own cellular material.

Yet grazers (like cows and termites) lack the digestive enzymes to turn the cellulose into a usable form. They rely on microbes—the bacteria and protozoa that live inside their digestive tracts—to do the work for them. In fact, microbes are just about the only living creatures that can digest cellulose and convert it back into energy and glucose. Without these microbes, none of us would be able to enjoy a sirloin steak or a glass of milk, because there wouldn't be any cows. They would soon starve to death on the small amount of nutrition available from the non-cellulose parts of the plants they eat.

From here, life gets more complicated and the carbon can go in many different directions. It may end up as food for a predator that eats the grazer, which in turn may be eaten by an even larger predator. The plant, the grazer, or the predator may die, after which it will be attacked by bacteria and fungi that specialize in decomposition. These microbes break down the organic material with its stored carbon and energy into smaller and smaller pieces, making both available for other members of the food web to use.

Methanogen

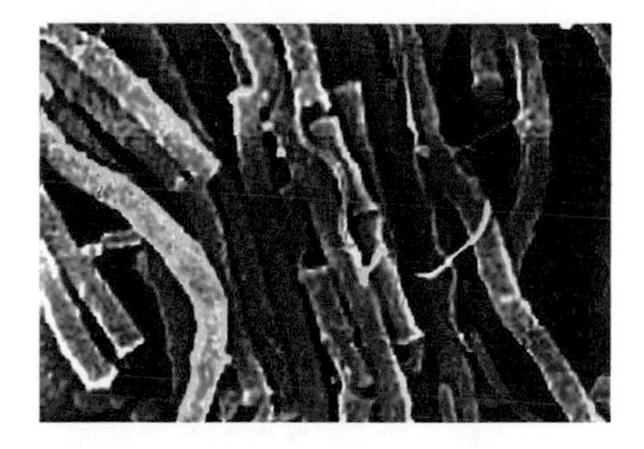

Identity: Bacterium
Residence: Insect guts
Favorite pastime: "Rumen"ating
Activities: A member of The Methane Makers, microbes like this live in the rumen of cows and inside termites, taking advantage of the waste products of others to generate natural gas (methane).

The termite's impressive capacity to get its carbon and energy from wood depends on the microbes living within it breaking down the cellulose in the wood. This is one way in which living things make use of the energy stored during photosynthesis.

Pedomicrobium

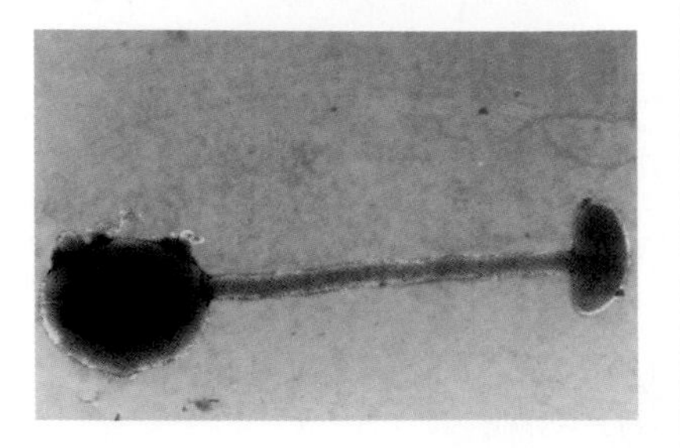

Identity: Bacterium
Residence: Streams
Favorite pastime: Creating rust
Activities: A member of The Mineral Eaters, *Pedomicrobium* can be spotted along the edges of streams, where it deposits iron oxides in large quantities, making the stream bank appear rusted.

Anywhere along the way, a living creature may break the food completely down to get all the energy from it in another cell process called respiration. This is a critical step in the food web and the carbon cycle, because the end result of respiration is the return of carbon to the atmosphere as carbon dioxide.

To understand the importance of carbon "recycling," imagine what might happen if some step in this process were stopped. If, for example, microbes were not around to decompose dead plants and animals, pretty soon a lot of the available carbon would be tied up in dead bodies. Plants and some microbes would continue to take carbon dioxide from the air, but less and less would be returned.

Ultimately, if all of the carbon were tied up in organic material and rock, the cycle might stop completely and life as we know it would cease to exist. There would be no carbon available to allow living organisms to build the materials they need to grow and reproduce.

The death and decay of a tree mean life to the bacteria and fungi that specialize in decomposing its substance.

The Great Geritol Experiment

Carbon dioxide is one of the important greenhouse gases. Its concentration in the atmosphere has increased over the past 200 years as a result of human activities. Scientists are becoming convinced that the increase in carbon dioxide is contributing to global climate change. Calculating its impact and finding ways to reduce the likely adverse effects of climate change—at least to us humans—presents scientists with a difficult challenge.

Scientists examining the ocean's microbial populations came up with the hypothesis that increasing the ocean's vast floating photosynthetic community might draw more carbon dioxide out of the atmosphere. They speculated that if they could encourage significant enough increases in their number, the microbes might remove enough carbon dioxide in building their cell material to offset the added amount that human activities are contributing.

An important limiting factor for the expansion of the photosynthesizing microbes in the ocean is iron. Iron is required in only tiny amounts, but it is a critical mineral nutrient for cell growth. It is present in a very low concentration in the open ocean. The scientists hypothesized that raising the iron concentration in the seawater would prompt an increase in the numbers of microbes carrying out photosynthesis and consequently a decline in carbon dioxide.

In 1993, the investigators tested their hypothesis in the warm waters of the equatorial Pacific Ocean, south of the Galapagos Islands. Here they distributed 480 kilograms of iron sulfate over an 8-square-kilometer area of the ocean. The test was dubbed the Geritol experiment, after an iron tonic marketed for senior citizens. Sure enough, the iron addition did what the scientists predicted—it stimulated the growth of the photosynthetic microbes, which doubled in numbers over the next three days. With the doubling of the microbes, the carbon dioxide in the surrounding atmosphere decreased.

The effect, alas, was short-lived. The added iron rapidly precipitated out of the water and fell to the bottom of the ocean, well out of reach of the carbon dioxide-consuming microbes. It's not likely that this particular strategy will allow us to reverse the greenhouse effect. But it does remind us of the power of the microbial world to cycle elements in a very big way.

Follow the Bouncing Carbon

floats over a leaf

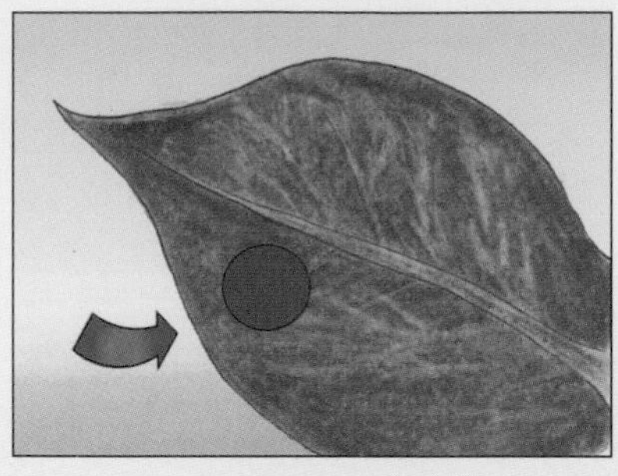

becomes part of a leaf

is transported underground

If we could attach a big blue dot to an individual carbon atom, we could follow its movements around the carbon cycle. Most of us would likely be very much astounded by the routes it took. We might find our blue dot, with its two atoms of oxygen attached, bouncing around in the air above the rainforest in Corcovada, Costa Rica, where it has been wafting up and down for years. On the day we discover it, our blue dot has just made its way into the little opening in the leaf of a *Cecropia.*

The next time we see it, the blue dot is still on the leaf surface, stored in the leaf's cellulose, but now the leaf is a fragment marching across the forest floor amid a parade of similar fragments. On closer inspection, the blue dot and its leaf are being carried on the shoulders of a well-known denizen of the forest—a leaf-cutter ant. The ant and its colleagues, all similarly laden, are returning to the deep subterranean chambers of their nest. There, more waiting ants chew up the leaf and blue dot, mixing it with ant droppings and depositing it in a rich mulch of similar material.

Growing on the mulch is a kind of fungus that lives with the leaf-cutter ants. In fact, the fungus and the ant are codependent, neither being able to exist without the other. The fungus degrades the leaf, sucking in the blue dot carbon and incorporating it into its cellular material. Several weeks later, the blue dot will return to the ant, this time in the form of a midnight snack. The ants cultivate the fungus just as we might a vegetable garden, and the fungus—along with our carbon—is eaten and digested.

The ant converts the fungal material into glucose

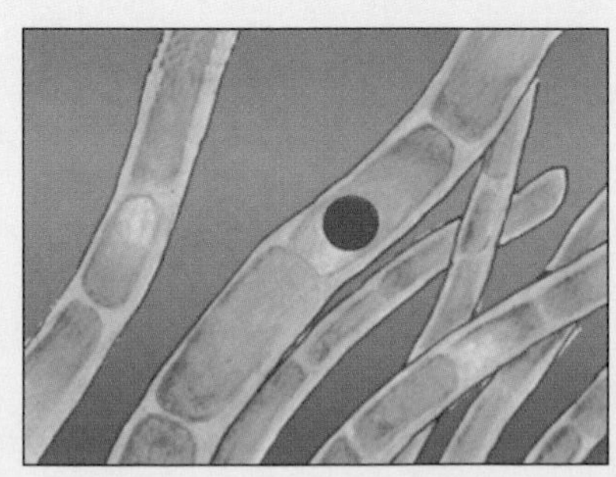

is eaten by another microbe, released as CO_2

becomes part of another alga

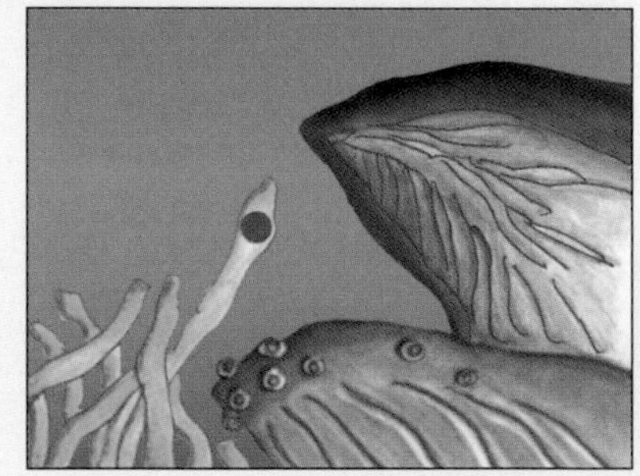

is eaten by a passing whale

is released to the atmosphere as CO_2

returns to float over a forest

is chewed up

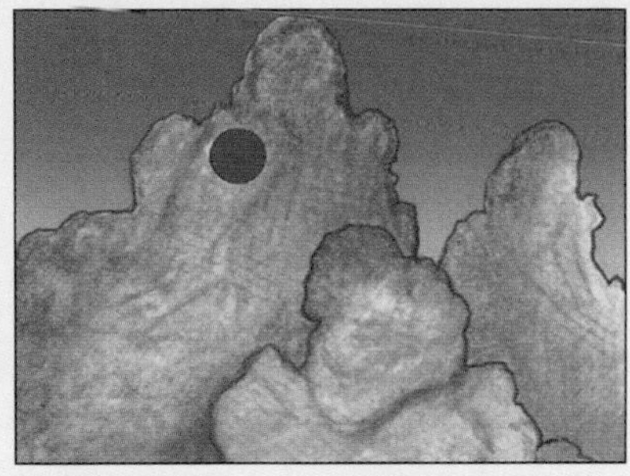

becomes part of a fungus

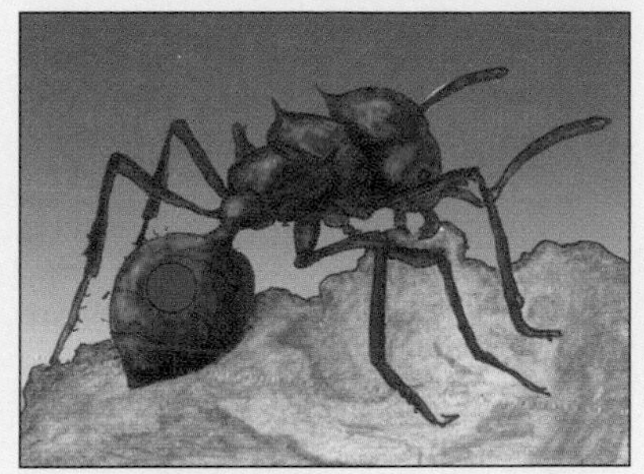

is eaten by an ant

and stores it for the next day's foraging march. A cell in the ant calls forth the blue dot special, taking oxygen back from the atmosphere, oxidizing the glucose (using oxygen's energy to separate the atoms in the glucose molecule), and converting the carbon atom back into carbon dioxide, which returns to the atmosphere. The wind pushes the blue dot carbon and its two oxygen atom companions out over the Pacific Ocean, where they dissolve into the warm, salty water.

is released as CO_2

The next stop is inside a large algal cell floating close to the ocean surface. The blue dot is once again freed from its companion oxygen atoms and enters the world of the animate as a product of the alga's photosynthesis. This time it is converted into sugar and incorporated into the alga's tough cell wall. The following week, a small copepod swims by, scooping up algae for dinner along with our blue dot. The copepod digests the algal cell, but not its cell wall, which it excretes with our carbon atom in its fecal pellet.

dissolves in the ocean

The fecal pellet drops slowly down, through 500 feet of water. As it drops, the blue dot is separated from the other carbon atoms in the fecal pellet. Microbes that shuttle from the ocean floor to the surface consume the carbon, releasing it back into the water as carbon dioxide, where it is again captured by an alga. This time a passing whale strains the alga from the seawater. The whale uses the dot as part of its respiratory process, returning the dot to the atmosphere once more as carbon dioxide.

A sea breeze sweeps the dot back to the rainforest, closing this simple version of the carbon cycle.

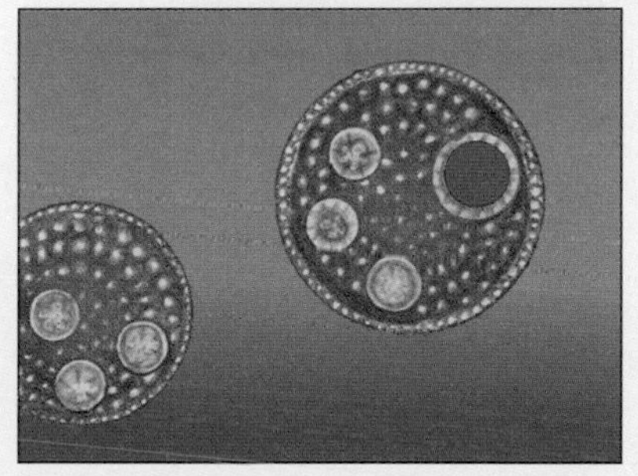

becomes part of an alga's cellulose

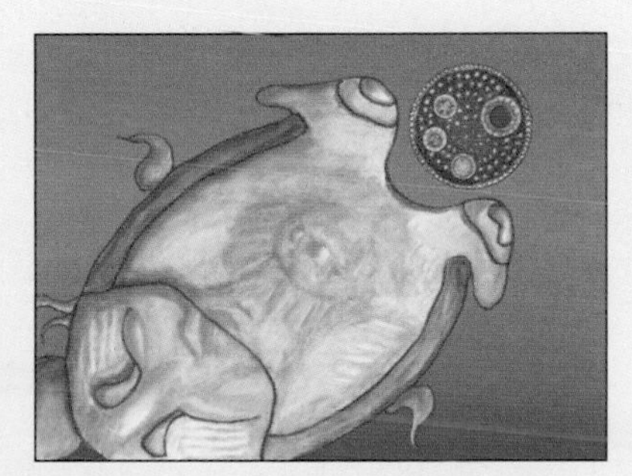

is eaten by a copepod

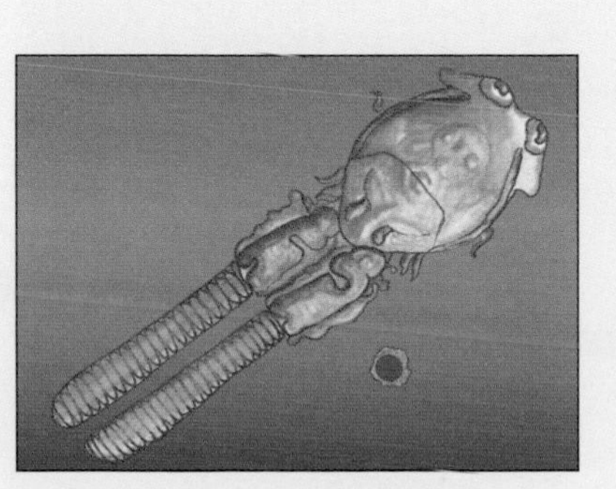

is excreted as cellulose waste

Cycling at Its Best

There are many cycles that sustain life in addition to the carbon cycle. The other major elements in living matter—hydrogen, oxygen, and nitrogen, along with sulfur and phosphorus—must also be cycled.

Take nitrogen. This element is a part of all living cells' DNA and proteins and is the fourth most common cellular component after carbon, oxygen, and hydrogen. The principal inorganic source for nitrogen, like carbon, is the atmosphere. In fact, over 99.9% of the nitrogen in our biosphere that is not tied up in rock or dissolved in water exists in the form of nitrogen gas (N_2) in the atmosphere. This means that less than 0.1% of the planet's nitrogen is used in—or readily available to—all living cells.

While either plants or microbes can capture atmospheric carbon, microbes are the key players when it comes to nitrogen. Plants, and consequently all the rest of earth's living creatures, depend on the microbes' ability to capture nitrogen from the atmosphere and convert it into a form the plants can use.

This dependence has spawned some unusual alliances. If you are wandering through a field or lawn filled with fragrant clover, pull up a plant and look carefully at the roots. Clover roots typically have an array of small nodules. The nodules are the visible imprints of one such plant-microbe alliance.

The clover nodules contain specialized microbes that make nitrogen available for the plant. The microbes take nitrogen from the atmosphere and convert it into ammonia, which then supplies both plant and microbe with a usable form of nitrogen for growth. The microbe's actions are called nitrogen fixation, and the groups of microbes that can carry this out are called nitrogen fixers.

Although some nitrogen is made available when lightning converts atmospheric nitrogen to a biologically usable form, microbes and humans have the exclusive biologic patents on this process. We build factories that produce synthetic nitrogen fertilizer by using chemical processes to fix atmospheric nitrogen. Synthetic nitrogen fertilizer is one of the mainstays of farming and gardening, allowing us to increase the yield of some of our most important agricultural products, such as corn.

Rhizobium

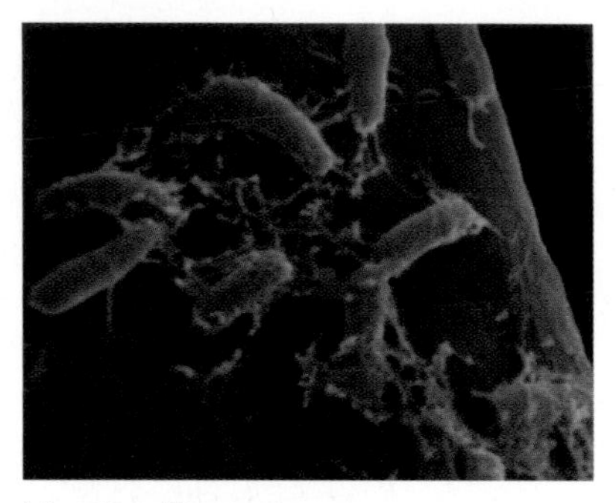

Identity: Bacterium
Residence: The root hairs of peas
Favorite pastime: Sharing food with its plant
Activities: A member of The Nitrogen Grabbers, *Rhizobium* captures nitrogen from the atmosphere and converts it into a form that both it and its plant can use.

Aside from the use of synthetic nitrogen fertilizer in human agricultural practices, however, all living plants depend primarily on microbes to supply their nitrogen requirements. In exchange, the plant supplies the microbe with nutrients for its growth, forming a perfect partnership. These plant and microbial partnerships are one of the most important alliances in the plant world. And, since we humans are entirely dependent on plants—and things that eat plants—for our existence, they are also one of the most important alliances in our world.

Mutual aid

A symbiotic trade-off: plants get a usable form of nitrogen, microbes get food.

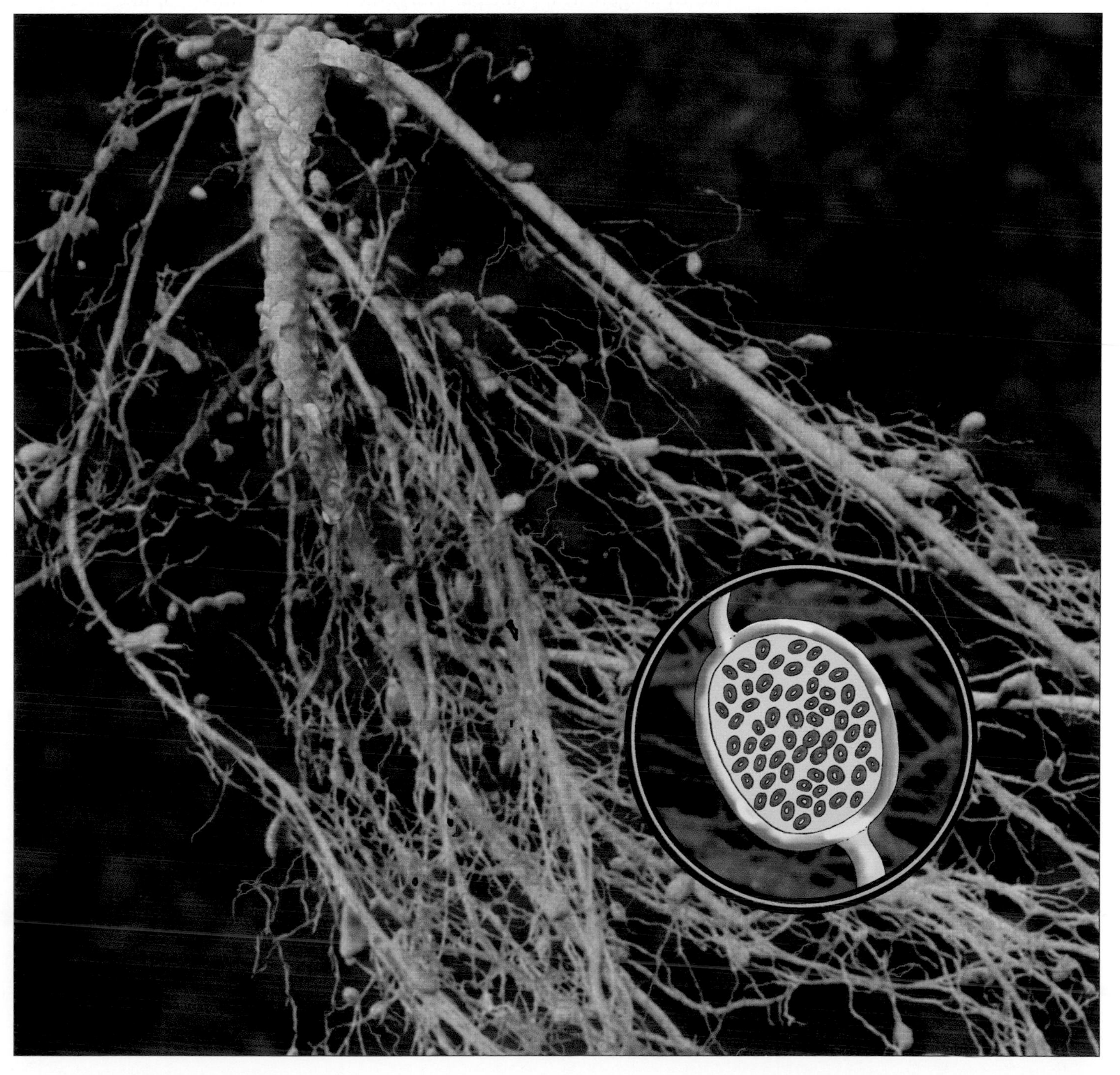

Perfect Partnerships

Such partnerships must have existed from the time land-based plants began to evolve. There is nothing random about the partnerships either, since particular plants associate only with particular microbes. In fact, plants and their specific nitrogen-fixing microbes are so attuned to each other that they have evolved a unique series of chemical communications to let each partner convey information to the other. Such chemical conversations are common throughout nature, including the ones between microbes and humans.

Consider the pea plant. As soon as a new pea plant emerges from its seed and begins to send out roots, it transmits a chemical message into the surrounding soil. If we could listen in, we might hear something like, "Come and join me; I'll trade you a safe home and food for a little bit of that extra nitrogen stuff you need to get rid of." Only the pea's particular microbial partners, in this case a bacterium named *Rhizobium,* pick up the message.

The pea's bacterial nitrogen fixers, picking up the welcoming message, propel themselves closer to the plant's tiny root hairs, sending their own message back to the plant. In a seeming display of affection, the root hair begins to curl around the bacteria, softening the tough walls of its cells to welcome the bacteria inside. Once the bacteria are inside the root's cells, the plant builds the equivalent of a microbial people mover, transporting the rhizobia to the very heart of the root.

The dynamics of cooperation

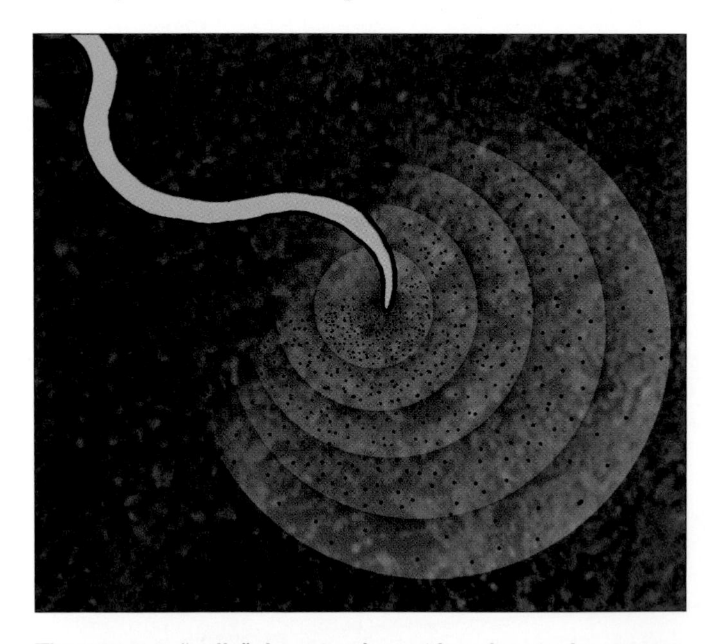

The pea root "calls" the microbes with a chemical message.

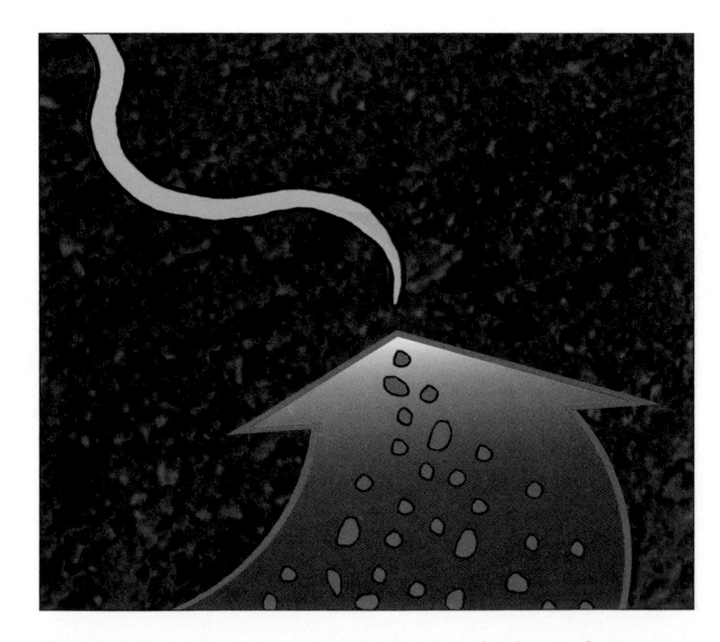

The rhizobia migrate toward the root and release their own chemical message.

The bacteria signal the plant's cells to acknowledge that they have arrived and begin taking nitrogen from the atmosphere, converting it into a form that both the pea plant and the bacteria can use. We can appreciate the success of the alliance by the luxurious growth of the pea plant and the plethora of little nodules filled with bacteria that populate the plant's roots. Without them, the pea plant would be small and sickly unless provided with nitrogen from some other source.

Legumes—like peas, beans, alfalfa, and clover—along with their nitrogen-fixing microbes represent one of the most important ways that nitrogen can be restored to land under intensive cultivation. Their ability to convert huge quantities of nitrogen to a biologically usable form, in fact, is the principal reason they are used in crop rotation. Their nitrogen-rich plant material can be plowed under, providing a source of usable nitrogen for crops like wheat, rice, and corn that lack this microbial partnership. By rotating the two kinds of crops, farmers are able to reduce their use of expensive nitrogen fertilizer. Without such rotation, the farmers must rely on human sources of nitrogen.

Microbes carry out a similar process in aquatic habitats. In this case, the prime cyclers are bacteria named *Cyanobacterium.* They are the only living creatures known that can both fix nitrogen and perform photosynthesis. Since they can perform both of these critical functions, they are the only totally self-sufficient living creatures—and among the most ancient. Microbes that look like today's cyanobacteria appear to have been the dominant forms of life for nearly 4 billion years.

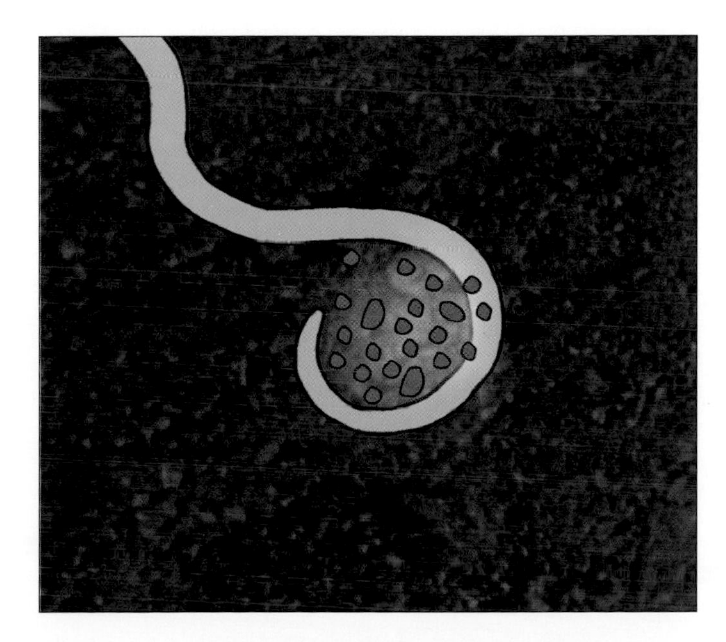

The root surrounds and engulfs the eager bacteria . . .

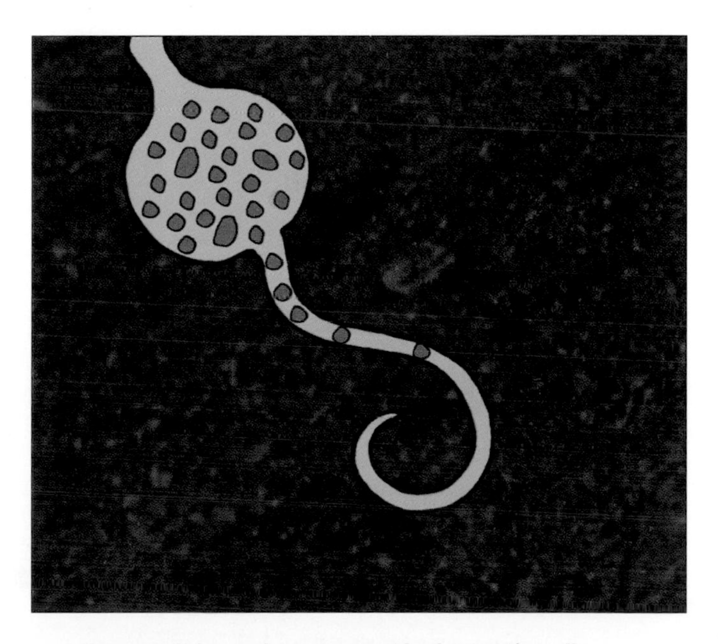

. . . who set up housekeeping inside the root's cells.

What Comes Around Goes Around

Just like carbon, once nitrogen enters the living world, it can follow a variety of routes on its way back to the atmosphere. Plants, animals, and microbes can all use ammonia—nitrogen atoms coupled with four hydrogen atoms—to build cell material (organic matter). After any living creature dies, microbes decompose the dead body, making the organic nitrogen directly available for reuse as ammonia. Other microbes can convert ammonia into water-soluble nitrogen salts, called nitrates, which can move freely in soil and water, providing another source for feeding the local inhabitants.

Aspergillus

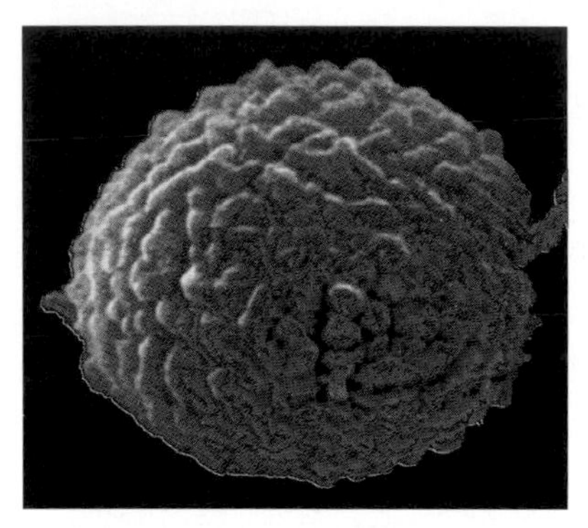

Identity: Fungus
Residence: Forest floors
Favorite pastime: Spoiling food
Activities: A member of The Decomposers, *Aspergillus* releases its digestive enzymes to break complex substances down into smaller ones; when this happens to our food we call it rot!

Of course, something must be transforming the nitrogen in biologic sources back into atmospheric nitrogen. If the story stopped with nitrogen fixation, microbes would have tied up all the available nitrogen in organic matter by now, where it would merely be traded around. It turns out that another major microbial force is at work, transforming organic nitrogen back into nitrogen gas. The microbes that manage this are so efficient, in fact, that they are partially responsible for keeping the amount of nitrogen in biologic form very low—less than 0.1% of the total nitrogen on earth. These so-called denitrifiers rely on various forms of nitrogen salts for energy, and with remarkable efficiency send nitrogen back into the air.

On the way back to nitrogen gas, microbes in the soil and water move the nitrogen atom through a series of different chemical compounds, some of which have important environmental implications. For example, an intermediary step in the cycle back to nitrogen gas involves the generation of nitrous oxide (sometimes called laughing gas). Nitrous oxide is a potent molecule that, when released into the atmosphere, depletes the ozone layer that shields the earth from the sun's damaging rays. Excessive nitrous oxide formation occurs when there is an elevated source of nitrate present in the environment, say when there is intensive use of man-made nitrogen fertilizer. Although there are many contributing factors in the depletion of ozone, there is little doubt that our extensive use of artificial fertilizer is among them.

And so it goes with all the materials that are used to build living cells. Complex communities of microbes are at work, constantly building, breaking down, and exchanging all the chemical compounds necessary for life to exist. A single tablespoon of fertile soil has more than one billion microbes in it, representing at least 5,000 different kinds, each playing a slightly different role in their microscopic community and each, in some way, linked to the cycling of materials. A teaspoon of seawater contains literally millions of microbes, each playing its part in keeping our biosphere in balance.

Microbes connect every living thing on the planet in neverending cycles, forming a complex web of life. They conduct millions of exchanges and are woven into the multi-dimensional relationship that surrounds and supports us all.

Decay is a process in which fungi and bacteria feast on the complex, nutritious arrangements of molecules that life has laboriously assembled. The enzymes within the microbes break these molecules down to smaller, reusable components, which are then built back up in the microbes' cells. In the process, oxygen is consumed and carbon dioxide and water are released. Thanks to this process, death yields life.

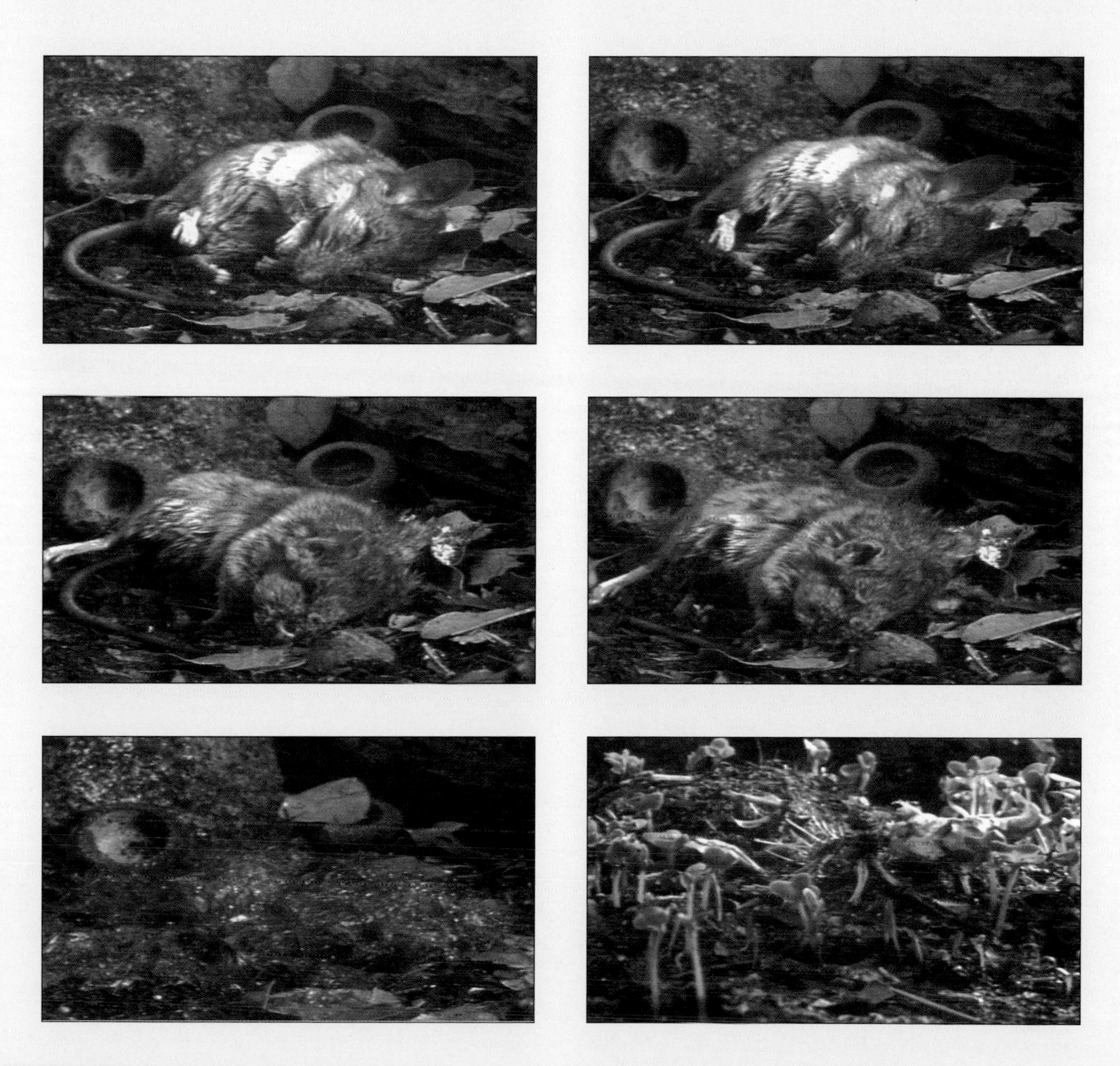

Dan Jantzen: ". . . When somebody says web of life we think two-dimensional—spider web . . . —but it's really a three-dimensional structure . . . the microbes all intertwined . . . think of them more as glue . . . sort of sticky everywhere."

Into the Rainforest

Nowhere is the diversity of life and its chemical exchanges more dramatic than in the lush tropical rainforests that encircle our planet. These rich ecosystems demonstrate an intensity of life that envelops visitors, overwhelming the senses and offering up new perspectives with each turn.

Members of the visible world abound here. A single tree is likely to be home to hundreds of insects, animals, and other plants. And for each member of the visible world, there are thousands of microbial species, intertwined in a humming enterprise whose complexity scientists have only just begun to dissect.

Costa Rica, the small Central American country that forms the narrow land bridge between the northern and southern hemispheres, has come to value its tropical ecosystems in ways not yet typical of developing countries. Its citizens have formed partnerships with leading scientists, creating natural laboratories for exploring the complex relationships that make such diversity possible.

Winny Halwachs: "I knew that [microbes] were out there but the sense that they had really structured the world around us, that they are the engineers that made life livable on planet earth and that they are the engineers that maintain conditions within the range that we can survive—that has come to me much, much more recently."

Biologists and microbiologists come to study in the natural laboratories of Costa Rica for different reasons. Dan Jantzen and Winny Halwachs, a husband-and-wife team, are in the process of cataloging as many living creatures as they can in the rich ecosystems that define Costa Rica. They hope that their catalog will help scientists gain a better understanding of the diversity that characterizes this region and to monitor future changes in its composition.

Jantzen and Halwachs have devoted their lives to restoring a part of the rainforests that were burned to make way for cattle in the northwest corner of the country called Guanacasta. Although not trained as microbiologists, they appreciate the role that microbes play in this specialized ecosystem. Jantzen describes them as the glue that holds everything else together. "When most people think of a web of life, they think of a two-dimensional structure. In reality, the web of life is multi-dimensional, and the microbes are intertwined throughout all of it."

Ignacio Chapella is another scientist who comes to the rainforests of Costa Rica to study. He works with a particular category of fungi that play an essential role in the rainforest's health and well-being. He is learning about the partnerships between fungi and plants that support the luxurious growth of the rainforest. Chapella has crossed paths with Jantzen and Halwachs because what Chapella is learning is helping them decipher and catalog the relationships among many of the diverse species in these important ecosystems in the tropical web of life.

Ignacio Chapella of the University of California: "I think the average person should . . . be interested in microbes because it's always good to know who is running your world . . ."

In the rainforest, materials descend from the canopy, are rapidly decomposed, and are swept back up in new growth. In contrast to temperate forests, the topsoil is little enriched because the copious daily rains wash any remaining nutrients away.

Decay among the Trees

Fungi play a significant role in keeping the mineral and nutrient cycles going in all terrestrial ecosystems, but they are particularly important members of the web of life in the tropics. Without fungi, the lush tropical rainforests would become barren wastelands.

Looking at the amazing plant growth in a tropical rainforest, one might naturally assume that the soil supporting the forest is rich and fertile. However, looks in this case are quite deceiving. The soil is actually nutrient depleted.

The high temperatures in the tropical climate cause decomposition of dead biologic matter to run at a far more rapid rate than in the more temperate climates further north and south. Thus nutrients are released much more rapidly into the soil on the rainforest floor. You would expect such rapid decomposition to create a fertile environment. But daily rainfalls quickly wash away any free nutrients. The rain leaves the soil leached of the materials necessary for plant growth.

When farmers attempt to convert rainforest sites to agricultural use, the effects of the daily rain on soil nutrients become quickly apparent. After the forest has been cut and burned and then planted with an agriculturally desirable crop, the soil becomes barren and often erodes in as little as a single season. The only way to compensate for this is to use substantial amounts of synthetic fertilizer, an expensive solution that is often beyond the means of farmers in developing countries.

So how, then, does the rainforest soil support such luxurious natural plant growth, while farming yields such paltry results? One secret rests in key partnerships between the plants and a group of fungi called mycorrhizae—literally, "fungus roots." Tropical plants and their fungus roots form a closed nutrient cycle that is impervious to the harsh effects of the tropical climate on nutrient availability.

These partnerships are so intimate that the fungi actually become a part of the physical structure of the plant root; hence their name. In some cases the fungi form a thick outer envelope around the roots and physically replace the root hairs, serving in their stead to draw nutrients from a large expanse around the plant. In other cases, the fungi invade the living cells of the roots, forming tightly wound structures that look like living brocade in the fabric of the root cells' interiors.

The plant and the mycorrhizae each clearly benefit from their partnership. The fungus roots serve as nutrient scavengers. As dead material falls to the forest floor, many different kinds of fungi and bacteria participate in its decomposition, breaking the organic material into smaller and smaller pieces for reuse. Mycorrhizae help both to decompose material and to quickly sequester key elements like phosphorus and nitrogen inside their own cells.

Such rapid sequestration traps the nutrients before the rain leaches them from the soil. The fungus roots then transfer the nutrients directly to the plant root. In exchange, the plant provides a safe haven from the mycorrhizae's fungal and bacterial predators, as well as additional nutrients, such as glucose and vitamins, that the fungus needs to grow. It is reasonable to suspect that these plant-fungus partnerships also evolved with the first land plants, giving them a way to marshal the nutrients necessary for their success. Today, 98% of all terrestrial plants have some sort of mycorrhizal partnership.

The fungus mycorrhizae wrap themselves around tree roots, essentially taking over the root's function, trapping nutrients for use by both plant and fungus.

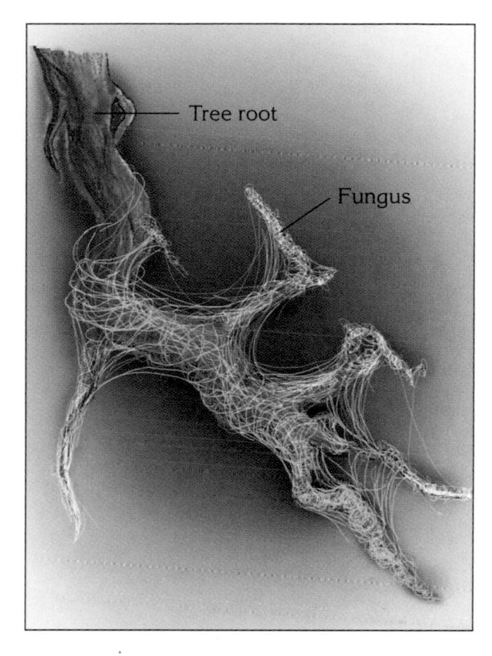

Mushrooms are communities of fungal cells specialized in the production of spores. They participate in the process that returns essential nutrients to drive life's cycles.

The Threads of Success

Although fungi are microbes, they have some unique physical characteristics that separate them from the predominantly single-celled existence of the rest of the microbial world. For one thing, they form long threads, filaments of living cells attached end to end, that can stretch across long distances. Because there are so many cells linked together, they often become visible to the naked eye. Leaving a piece of freshly baked bread out on the counter in the kitchen for a couple of days is usually a good way to demonstrate a collection of fungal threads. We recognize them in this case as bread mold!

These threads are called hyphae. They can stretch for many kilometers through the soil, with each cell in the thread able to communicate and pass nutrients and minerals to other cells along the extended chain. When the fungi are root fungi, they can also transport materials needed for growth from one plant to another. From a plant's perspective, this provides the equivalent of an underground superhighway for moving certain critical

Master movers

The fungus's long chains of connected cells—hyphae—act as conveyor belts for nutrients throughout the forest floor.

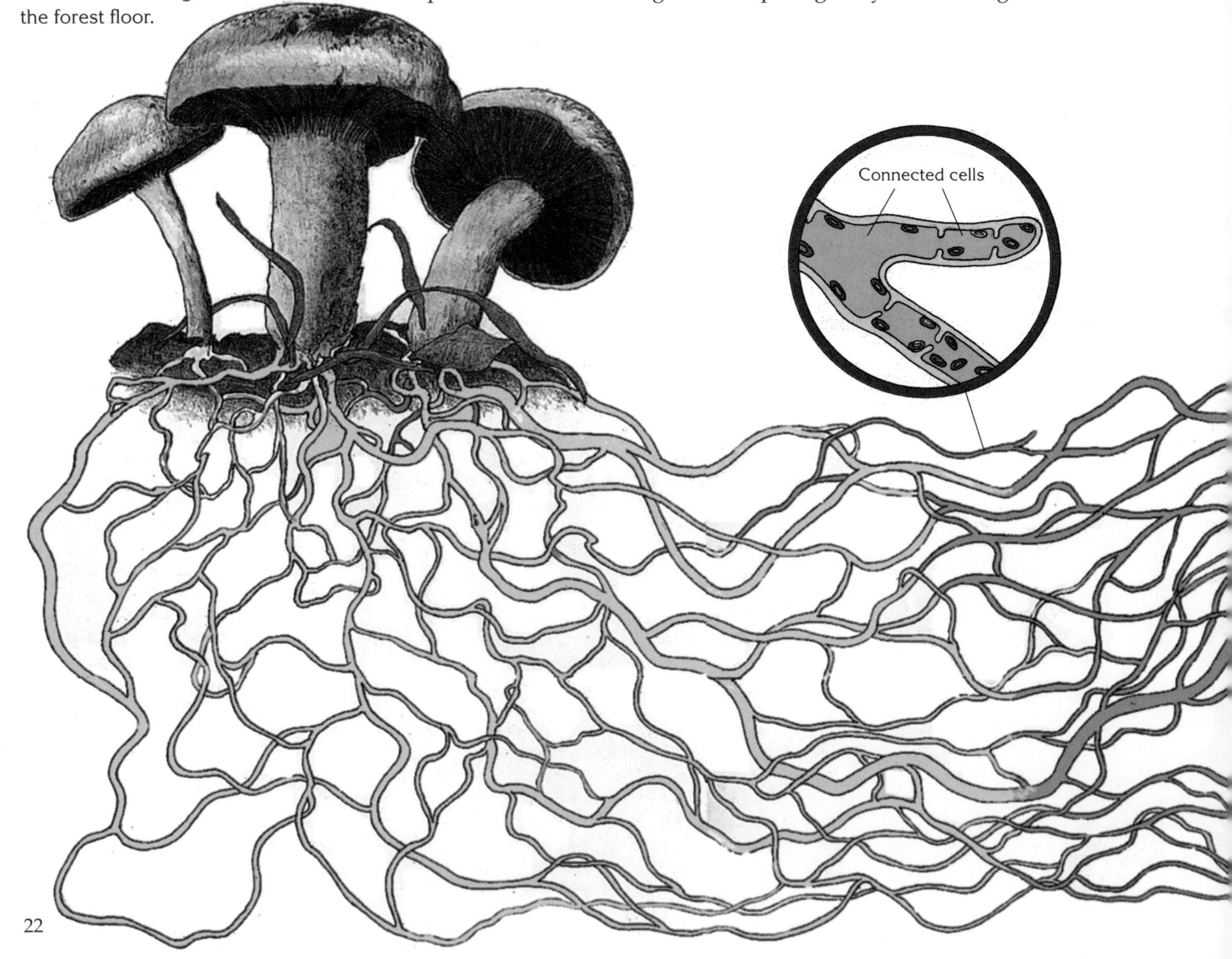

nutrients, like phosphorus and nitrogen, from areas of higher to areas of lower concentration. Thus, fungi play a very important role in helping terrestrial plants gain vital resources while the plants remain stationary, securely rooted to the ground.

Fungi that form threads are also proficient at penetrating between the cells of dead organisms. This makes them well suited to their role as the primary decomposers in dry terrestrial ecosystems, where they can't rely on the presence of water to move easily to the next meal. They simply grow toward it. Such is not the case in aquatic ecosystems where single-celled microbes, usually gifted with massive propeller power, function more efficiently.

Their ability to penetrate between cells doesn't stop, however, with dead organisms. Certain fungi are among the most important pathogens of plants, using their long threads to invade the cells of living organisms, and beginning the process of decomposition long before their victim is ready for that step in its life cycle. Fungi also prey on insects, other fungi, and animals, including humans.

Leaf-cutter ants, prodigious weight-lifters, slice off chunks of leaves . . .

From Forest Canopy to the Stomachs of Millions

. . . and transport them to their underground nests, where the leaves are food for a fungus, and the fungus is food for the ants.

There are other fungal partnerships in the rainforest ecosystems that are important in transferring energy and nutrients within the web of life. In a mature forest, almost all photosynthesis is conducted high in the treetops, in the so-called forest canopy. This is because sunlight can't penetrate much below the dense topmost layers of leaves and branches. In fact, the more mature the forest, the less plant life exists below the leaves at the tops of the tall trees.

Since the majority of the energy stores are overhead, it is important for the forest's diverse populations of living creatures to find a way to reach this resource. One significant way in which this vast biologic storehouse of energy is transferred to other living creatures is through another ancient partnership.

Leaf-cutter ants are responsible for transferring as much as a third of the carbon-based energy stored by the leaves in the canopy to the forest floor. The critical link in the transfer of the stored energy, however, relies on more than just the ant. In fact, the ants can't digest the leaves they carry from the canopy at all. The ants deliver the leaf fragments to their nest underground, where they chew them into tiny bits and place them in their vast subterranean garden.

The ants are bringing the leaves to the fungus they farm. The fungus decomposes the leaves and makes the nutrients and energy available to the ants as well as to other local inhabitants, both visible and invisible. The fungus depends on the ants to keep predators at bay and the food supply steady. The ants even supply critical enzymes that aid the fungus in digesting the leaves.

The ants have carefully cultivated this fungus, much like horticulturists might cultivate a prize strain of a particular rose. In fact, over millions of years of evolution the ants have passed along their fungal partner to each new leaf-cutter nest, so that the very same genetically identical strain of fungus exists wherever there are leaf-cutter ants.

The ants even guard their fungal farms against the encroachment of fungal predators, including a fungus called *Escovopsis* that is found nowhere except in the ants' nests. To do this, they have turned to another microbial partnership with a bacterium known well to the pharmaceutical industry—*Streptomyces*. These bacteria produce a variety of important antibiotics, and the ants have capitalized upon this trait. The ants carry the bacteria on their thorax and use the antibiotic they produce to kill the marauding *Escovopsis*.

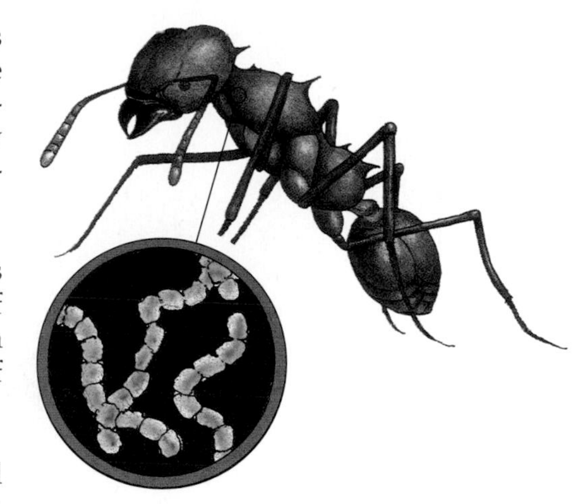

The leaf-cutter ants carry a special kind of bacteria, *Streptomyces*, on a patch on their chests. The bacteria, in exchange for this safe haven, produce an antibiotic that kills a mold that attacks the fungus that feeds the ants.

The Cows of the Canopies

Another partnership important in transferring stored energy from the treetops is one that Ignacio Chapella is studying. He refers to the partners as the "cows of the canopies." Caterpillars, a stage in the development of butterflies and moths, are voracious eaters. Certain of them feast on the leaves of trees forming the canopy, transforming the photosynthetically derived energy stores into their rapidly growing bodies.

Scientists suspect that caterpillars harbor microbes that help them digest their food just as microbes help cows digest theirs.

Chapella now believes that these caterpillars, much like cows eating grass, cannot digest the leaves without help from a friend. He speculates that the caterpillars rely on the action of fungi that populate their gut to digest the leaves, converting their stored energy into a form usable for both the caterpillar and the fungus. From here, the converted nutrients can enter the food web in a variety of ways, moving carbon and other key elements from the treetops to the life forms below.

This is a partnership that, like the one between the cow and the microbes that populate its multiple stomachs, results in the transfer of energy. With the help of microbes, the cow thrives, converting the energy from the grass into meat, milk, and baby cattle. With the help of microbes, the caterpillar thrives, eventually becoming food for something else.

Phytoplankton, microbes inhabiting the upper, warmer, illuminated layers of the oceans, are the primary converters of sunlight to living substance in these watery expanses.

Oceans of Microbes

Most of us are at least somewhat familiar with the way life exists and interconnects on land. Life in the dark and mysterious oceans, however, from the giant squid to the tiniest microbial residents, remains one of earth's last unexplored biologic frontiers.

In at least one way, oceans bear a resemblance to tropical rainforests with their enormous life-supporting canopies. The upper layer of water is an active zone of primary food production. Like the rainforest canopy, the upper layer of water is where light from the sun can penetrate and photosynthesis occurs. Just as on land, this first step in transferring energy from the sun into the ocean's web of life is a critical step for the majority of the familiar, larger life forms that populate the upper reaches of the ocean.

There is a major difference, however, in who is responsible for primary food production. In the rainforest, as in all other land-based ecosystems, it's plants. In the ocean, primary food production is the exclusive province of microbes. Bacteria and algae—together known as phytoplankton—are responsible for energy capture via photosynthesis just like land-based plants. They live on or near the ocean surface, floating in the relative warmth where they have ready access to the sunlight they need.

At the ocean surface, complex communities of microscopic and macroscopic life interact, transferring and recycling materials necessary for growth and reproduction in the watery environment. It is a certainty that the same sorts of partnerships between microbes and larger life forms that exist on land are present and operating to keep life rolling along in the ocean. Microbes play as important a role in the cycling of materials among the ocean's residents as they do on our more familiar terra firma.

The Hunt for SAR 11

Giovannoni and his team collect samples in pursuit of the elusive SAR 11 in the Sargasso Sea off the coast of Bermuda.

Microbiologists like Steve Giovannoni and his students at Oregon State University are among the scientists hoping to learn more about the microscopic world that supports life in the oceans' various and remarkable ecosystems. The microbes that Giovannoni and his students discovered in their quest may prove to be one of earth's most abundant life forms. These microbes may also be among earth's smallest living cells.

Giovannoni's quest began over a decade ago. He, like others, knew that most of the vast numbers of kinds of microbes that exist have yet to be discovered. Scientists have been unable to grow them in the laboratory because they can't adequately replicate the complex communities and nutrients that support the microbes in their natural environments. He decided to apply a new technique developed by a colleague, Norman Pace, a technique that can identify microbes through their genetic signature. With Pace's approach, the microbes' unique sequences of DNA, life's information molecule, can be amplified and examined directly from the environment without ever having to grow the microbes in a laboratory. For his own studies, Giovannoni turned to the open ocean.

Steve Giovannoni, on the discovery of SAR 11: "My feeling was like being the first person to land on the moon."

"Clearly there was something special about this organism that allowed it to move in and occupy the biggest environment on the planet—the ocean's surface—where it then became the dominant organism."

The first time Giovannoni spotted his quarry was in samples of water he collected from the Sargasso Sea. The Sargasso is an area just north of Bermuda that typifies the conditions found in the open ocean. From the biologic standpoint, the waters are not particularly conducive to an abundance of life. For one thing, the nutrients necessary for growth and reproduction are very dilute. Despite this, Giovannoni detected literally millions of microbes in as little as a teaspoon of water from the Sargasso when he probed his samples using Pace's molecular techniques.

One kind of microbe accounted for at least a quarter of the cells present in the sample. The microbe's genetic signature didn't match any known microbe, and so the scientists dubbed it SAR 11 after the place where it was first discovered. Since that time, Giovannoni and others have found microbes with the same genetic signature in every body of water they have tested. SAR 11 and its relatives may well account for a quarter of all the biomass in water, which would make these microbes the most prevalent life form on earth!

What is this mysterious microbe doing out there? If it is so abundant, it must have a significant role in the complex communities in which it lives. What role is it playing in the tightly connected cycles that make life possible in the low-nutrient environment of the open ocean? A decade after SAR 11's discovery, these questions still remain unanswered.

What Giovannoni and his team have discovered is that the microbe seems to be well adapted to living with limited resources. Through the use of specially designed genetic probes, the researchers got SAR 11 to reveal itself under the microscope. Once visualized, the world's most abundant life form also turned out to be one of the world's smallest. SAR 11 is less than half the size of any other marine microbe, measuring in at just a little over 0.3 micrometer. Giovannoni speculates that the small, round shape of SAR 11 may expose a maximum surface area to sea water, allowing every opportunity for dilute nutrients to collide with and be taken into the tiny cell.

World's smallest bacterium?
This fuzzy dot of light, magnified by thousand of times, was the first confirmed sighting of SAR 11.

Giovannoni has hypothesized that SAR 11 exists in a tightly linked partnership with algae, one of the most important groups of primary food producers in aquatic ecosystems. If his speculation is correct, SAR 11 will be a critical link in the ocean's food web.

Much work lies ahead if Giovannoni is to prove his hypothesis. Not the least of the challenges is to coax the tiny microbe into growing in the laboratory, a time-honored strategy for gaining an understanding of how various microbes carry out their lives. If SAR 11 does share a tight connection with the ocean's algae, it will join the ranks of a few other well-known linkages among the ocean's varying ecosystems.

Wild Microbes Are Hard To Grow

SAR 11, like most other microbes in the wild, turned out to be a challenge to grow in a laboratory. Despite being tempted with many different foods, the tiny microbe could not be coaxed to multiply.

Researchers have not yet been able to identify the chemical and physical factors needed to reproduce its normal habitat, demonstrating how complex and interdependent microbial life can be.

Life in a Watery World

Although water may seem like just water to us, in reality the aquatic terrain varies geophysically every bit as much as the more familiar expanses of land do. Not surprisingly, the living arrangements within these different environments in the ocean are as distinctive and different as those found on land. Studying these ecosystems has offered some major surprises in terms of who is feeding whom.

Consider the coral reefs that abound in the warm, shallow areas of the ocean. They are luxurious underwater islands of highly diverse species all depending on each other for their existence. There are schools of brightly colored fishes. Crabs and eels hide amid the bizarrely shaped coral structures. There are sponges and sea anemones of all shapes and sizes. This richness of life is in contrast to the comparatively barren stretches typical of most of the ocean's terrain, where nutrients are too dilute to support such dense communities.

The entire reef is microbially powered. Even the physical characteristics of the reef are the direct consequence of a microbe-animal partnership. The coral itself is formed by a living animal with a hard, calcified exterior sometimes called an exoskeleton. The coral polyp depends on a partnership with a specialized kind of algae living in its tissues. Neither can live separately. The growth of the coral depends on food produced by the photosynthetic algae. The growth of the algae depends on a waste product, ammonia, excreted by the coral. Even the formation of the coral's hard skeleton is dependent on the removal of carbon dioxide from the water by the algae, shifting the delicate balance between two chemical forms of calcium carbonate.

Similar relationships exist throughout the reef. Certain clams, the anemones, sea fans, and sponges all contain photosynthetic algae and cyanobacteria within their tissues. Some of these animals even help their photosynthetic partners by moving into the best positions to bathe in the filtered sunlight, a condition that makes food production run more efficiently. These mutualistic relationships are so important that the health of the entire reef community is threatened when their exchanges of carbon, nitrogen, phosphorus, and hydrogen are interrupted.

Within the reef community, literally hundreds of thousands of exchanges of materials and energy go on continuously among all of the residents. All are linked in some way to the diverse microbial species that drive the cycles.

Many familiar inhabitants of the reef community and their unique microbial partners exchange materials that each needs to thrive.

Coral Woes

There is no more compelling way of illustrating the importance of microbial partnerships than through the lenses of the underwater camera. Coral is among the most magnificent of animals, with its heads ranging in color from soft green to dramatic fluorescent purple. Coral gets its color from the algae that live within its cells.

When the coral is stressed, however, it dispels its microscopic algal partners, leaving the coral's tissue translucent and revealing the white limestone skeleton. The coral, no longer fed by the resident algae, enters a starving stage where it can no longer grow or reproduce. If not reversed, the coral will eventually die. This condition turns the coral as pale as if it had been soaked in bleach, and researchers call it coral bleaching.

Coral bleaching is a general response to many different kinds of stresses, including ocean temperatures that are too cold or too hot, salt concentrations that are too high or too low, sunlight that is too intense or not intense enough, and suspended sediments that are excessive. When the stress is removed, the algae return, and the coral, if it has survived starvation, begins to thrive once again.

Ocean observers have witnessed coral bleaching for nearly a century, but before the 1980s the events were very limited in extent and duration. Beginning in the 1980s, however, bleaching spread dramatically, from the few small areas clearly affected by local conditions to huge areas of the ocean covering up to thousands of kilometers. The massive nature of these events set off alarm bells among coral reef biologists.

Scientists knew from laboratory studies that coral is extremely sensitive to temperature, and they suspected that temperature might be playing a role. Their suspicions were confirmed. The only stress factor that was universal to the mass bleachings was excessively high temperature. Wherever the sea surface was 1 degree Celsius or more above average in the warmest months of the year, the coral expelled its algae and bleached. Based on these observations, scientists fear that coral reefs are among the most temperature sensitive of all ecosystems, and consequently coral bleaching may serve as the best sentinel of global warming.

Recently, another phenomenon associated with ocean warming is creating perilous conditions for the world's coral beds. Previously unknown bacteria, fungi, and viruses appear to be attacking and killing coral even in the most pristine of waters. In January, 1997, coral researchers reported that a rapid, wasting disease was spreading across the coral heads at a rate of several inches a day in the reefs off the island of Bonaire. Since then, the disease has been spotted in Mexico, Aruba, Curacao, Trinidad, Tobago, Grenada, and St. John's in the Virgin Islands, an area spanning 2,000 miles.

More alarming is the fact that the wasting disease spotted in the Caribbean is only one among a variety of diseases discovered attacking coral around the world. Scientists are baffled by the diseases exploding on the reefs, but many fear that we are seeing the effects of rising sea temperatures in yet another form, making the coral and the delicate reef ecosystems they support much more susceptible to diseases they have escaped in the past.

Corals "bleach" (appear white) when they expel the algae that normally live inside them and provide them with most of their food. This forced separation occurs primarily as a result of rising tropical sea temperatures.

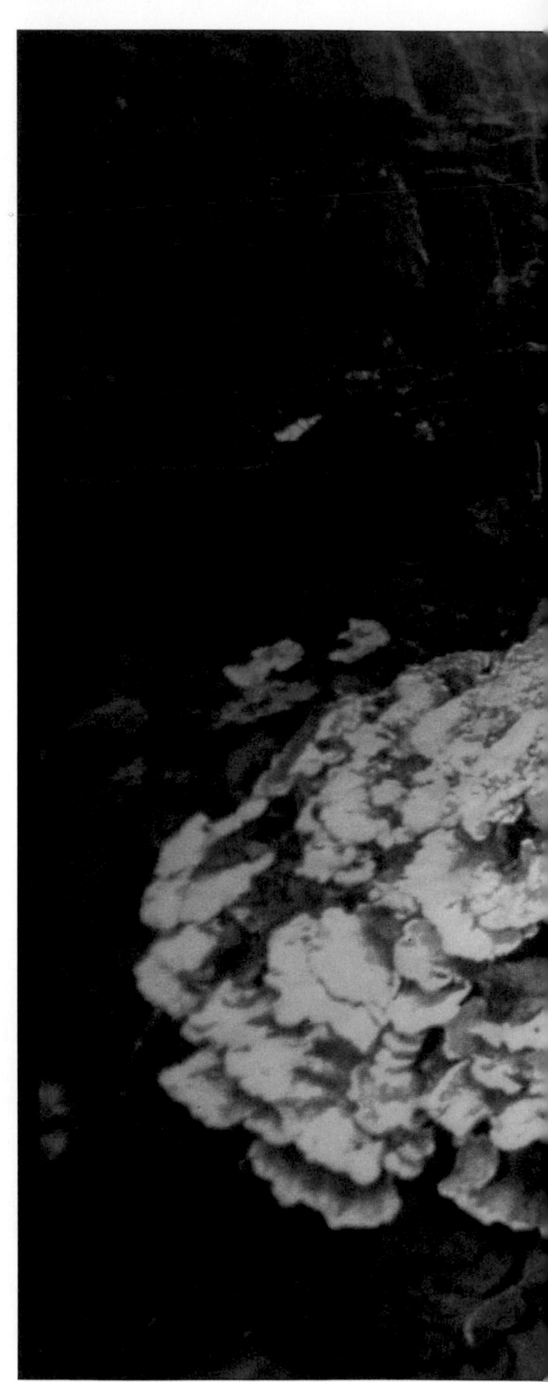

At the end of the millennium, almost 10% of the coral worldwide has died. If current trends and conditions continue, an additional 10 to 20% of the world's coral reefs could be lost. Although this may seem a small part of the reef system to lose, the reefs support significant fish populations, providing food for both people and the other inhabitants of the oceans. They also harbor enormous species diversity, some with the potential for yielding new pharmaceuticals and other commercially attractive products. Their loss diminishes the diversity of our planet and warns that our global environment is changing.

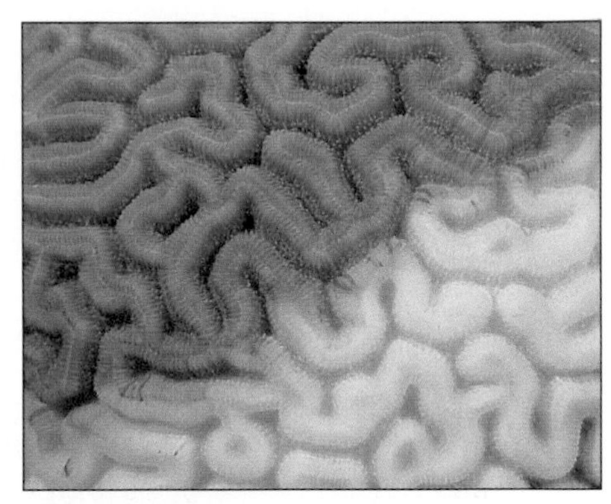

The disappearance of green algae from its coral host bodes ill for the reef.

Of Microbes, Sulfur, Clouds, and Climate

Even most microbiologists are surprised to learn that microbes play a direct role in the warming and cooling of the earth. This story is another reminder that, although tiny, microbes can collectively make big changes in our world.

The story began to take shape in the early 1970s when James Lovelock, a noted biologist and planetary ecologist, discovered an important link between microbes and sulfur—another essential element for living organisms. Like others, he knew that millions of tons of sulfur each year are washed into the ocean from land. He realized that some mechanism must be in place to return the sulfur to the atmosphere and hence back to life on land.

Lovelock and others discovered that marine algae produce a sulfur-containing compound called dimethyl sulfide, DMS for short, and emit it as a gas into the atmosphere. This algal waste product is especially prominent over the "deserts" of the open ocean, but even near land, DMS provides a bit of the aroma we associate with the sea. Once returned to the atmosphere, the compound combines with oxygen to form sulfuric acid droplets, which can then waft over the great land masses and be captured for use by terrestrial ecosystems.

But how does this have anything to do with climate? It turns out that the earth's temperature depends on how much heat is absorbed from the sun versus how much is reflected back out into space. Dark surfaces absorb heat (think about how warm a dark car becomes in the sun) and light surfaces reflect heat. On planet earth, the dark surfaces are the oceans and forests and the light surfaces are the ice caps, snow, and clouds.

Clouds form when water vapor from the atmosphere condenses on small particles in the air. These small particles are called condensation nuclei, and the very best material for forming condensation nuclei is a sulfuric acid particle. Clouds are exceedingly important in forming a reflective barrier that keeps the earth cool.

Now we get to the interesting part. When the sun heats the ocean surface, it causes the ever-present algae to "bloom," or increase in number. The presence of more algae greatly increases the amount of DMS released into the atmosphere (more algae, more waste products), and the DMS is rapidly oxidized to sulfuric acid. Clouds form around the nuclei, which reflect heat away from the ocean surface, allowing cooling of the surface waters. This, in turn, slows down the growth of the algae, which slows the rate of DMS release, which decreases the amount of cloud cover.

One of the cycles that balance earth's temperature is complete, and microbes are right in the middle of the action.

Microbes influence climate

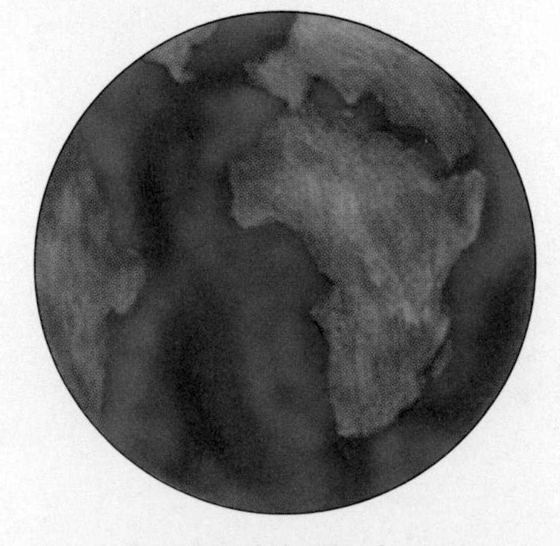

In the absence of clouds, the sun heats the ocean's surface and algae bloom.

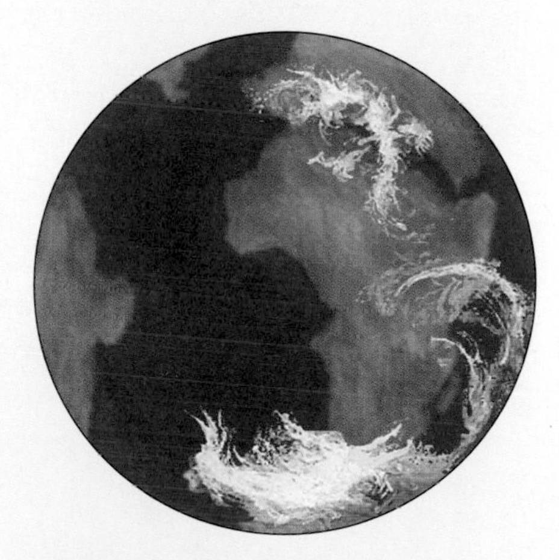

As algae multiply, they release more of the gas DMS into the atmosphere, causing clouds to form . . .

. . . which shade and cool the ocean, reducing the rate of algal growth and, thus, the release of DMS.

As DMS wanes, the clouds dissipate, the oceans warm, and the cycle begins anew.

Life in the Deep Sea Trenches

The lights of the submersible explorer ALVIN illuminate towering masses of crystallized minerals emanating from volcanoes two miles below the ocean's surface.

The human passion for exploration has fueled attempts to reach the summits of the highest mountain peaks and the bottoms of the deepest oceans. The deepest oceans, however, offer some challenges that the highest mountains don't. The pressure at 11,000 feet under the ocean surface is enormous, there is no light, and the temperatures are close to freezing. Very little organic material ever reaches these depths—creating conditions that could hardly be expected to support life. Scientists consequently assumed that the deep sea floor would be barren.

In the 1970s, two intrepid explorers from Woods Hole Oceanographic Institute (WHOI) were finally able to reach these depths. They were piloting a small, oblong vessel named ALVIN, a submersible specially designed to withstand the immense atmospheric pressure of the deepest ocean. Their goal was to explore a deep sea vent, one of the openings in the sea floor that allow geothermally heated water to exit.

And there a surprise awaited them. A rich and bizarre community of life forms never seen before—giant tube worms, unique clams and mussels, shrimp, and of course, microbes—spread in dense living mats over every available surface on and near the vents.

At the time, no one could understand how such communities could exist. No sunlight penetrated these depths, and nutrients were scarce. All other then-known ecosystems depended on photosynthesizing microbes and plants to link the food web together—impossible so deep beneath the ocean surface. Yet life flourished in the area surrounding the vent.

Another surprise awaited the scientists exploring this unique community.

Deep sea volcanoes support a thriving ecosystem of bizarre life forms, including ghostly crabs and dense mats of microbes.

Fire and Sulfur

The spreading of the sea floor from mid-ocean ridges creates rifts. The rifts, fractures in the floor, allow seawater to percolate into the earth's crust, where it comes in contact with hot basaltic rock. The water leaches various minerals from the rock, and the mineral-laden water moves back up through the sea floor and exits through vents, superheated and devoid of oxygen.

Movement of the earth's crust on the ocean's floor, as in an earthquake, creates cracks. Water flows in . . .

The water shooting up through the hydrothermal vents has certain unique characteristics. The heat of the earth's inner core has superheated it so that water temperatures in the middle of the vents significantly exceed the boiling point, reaching temperatures of 350 to 400 degrees Celsius. The water is laden with minerals such as iron, copper, manganese, and zinc sulfides, which precipitate when they hit the cold water above, creating the impression of smoke billowing from the openings.

Ecosystems like the one first discovered by the WHOI scientists surround the deep sea vents. The energy and elements that support these ecosystems are derived almost entirely from the water and dissolved minerals. Life in the deep sea vents exists because of microbes that "breathe" sulfur. In so doing, they capture the energy from the chemical bonds—much like capturing the energy from sunlight—couple it with carbon dioxide dissolved in the water, and transform it into food that supports all the other creatures in the community.

. . . heats up, then spews forth laden with minerals.

Novel relationships exist among the microbes and the larger inhabitants of these distant ecosystems. In some cases, the inhabitants simply graze on the sulfur-breathing microbial mats, much like sheep or cattle eating grass in a pasture. Thus, the microbes themselves serve as food, passing their nutrients and energy on to other life forms.

In other instances, the microbes exist in very intimate partnerships, almost becoming one with the larger creatures that depend on them for building materials and energy. Consider the vestimentiferans—the giant tube worms attached to the deep ocean bottom that peek shyly from their tough white outer cases. Although vestimentiferans cannot be grown outside their vent community, scientists have been able to piece together some of the features of their unusual life style.

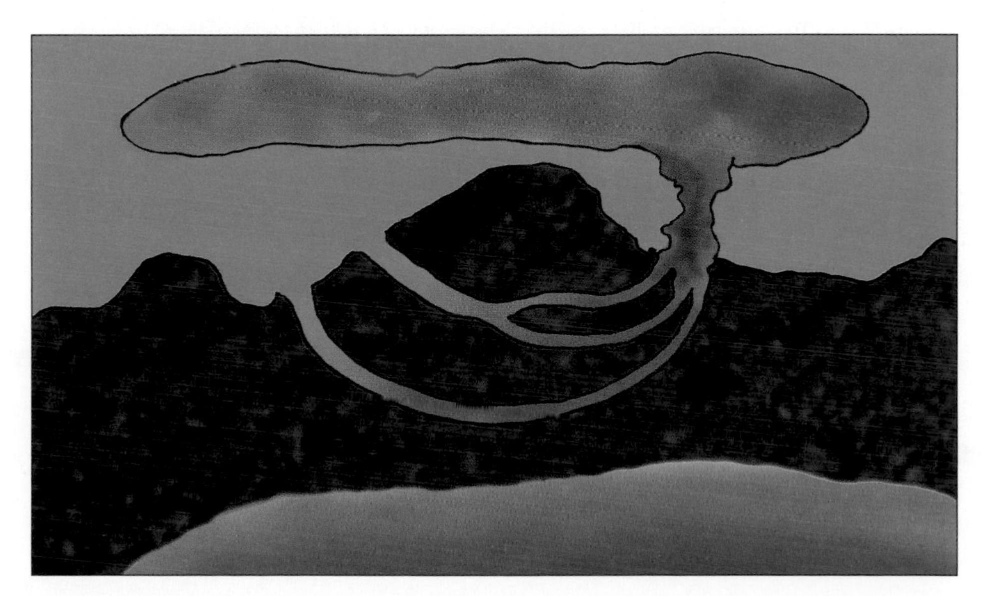

Microbes extract energy from the minerals, using it to convert carbon dioxide into their cell structure. They are, in turn, consumed by larger creatures.

In this sunless world, giant tube worms and other creatures depend on microbes to extract energy from the mineral-laden waters and convert it to a usable form for all.

The Life and Times of the Vestimentiferans

The vestimentiferan begins life as a tiny, free-swimming larva, complete with digestive tract. The larvae graze on the microbial mats surrounding the deep sea vents, perhaps capturing some of the microbes that will become a significant part of their lives later on. At some point in their tube worm adolescence, the larvae attach themselves to surfaces and mature into adults.

The elegant adult worm prospers in this environment, growing to upwards of 2 meters long. It is encased in a tough, whitish tube, into which it can retract for protection from predators. A brilliant, blood-red plume, the gills of the worm, waves from the top of the protective tube. The red color is due to the presence of a unique kind of hemoglobin molecule similar to the one in our red blood cells—with one very big distinction. The molecule picks up and transports hydrogen sulfide, along with oxygen and carbon dioxide, into the worm's tissues. (If we were to breathe hydrogen sulfide in such high concentrations, it would bind irreversibly to our hemoglobin. We would soon suffocate because our blood could no longer transport oxygen.)

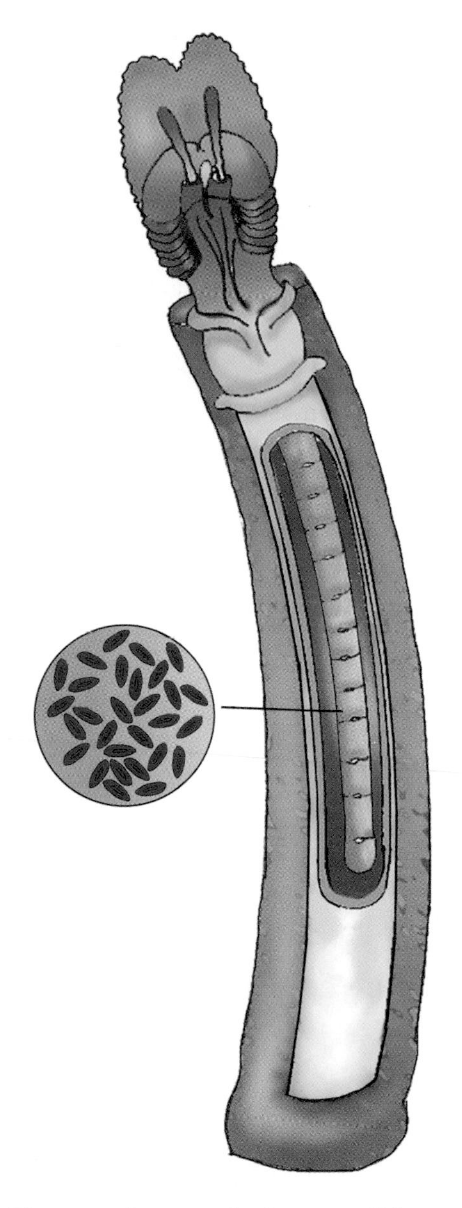

The worm harvests hydrogen sulfide, oxygen, and carbon dioxide from the sea and affords the bacteria a comfortable home; the bacteria convert the simple chemicals to food for themselves and their worm host.

Something strange happens to the tube worm as it reaches maturity. The worm's entire digestive tract disappears, including its mouth and anal vent. The worm is now a completely closed, self-contained unit.

So what can this giant worm with no mouth be living on? It turns out that over half of the tube worm's tissue harbors specialized bacteria. These bacteria extract energy from hydrogen sulfide, combine it with carbon dioxide, and create all the food supply necessary for themselves and their worm host. The tube worm exists entirely off the food produced by these tissue-dwelling microbes.

In exchange, the worm protects the bacteria, providing the right quantities of hydrogen sulfide, oxygen, and carbon dioxide for the bacteria to optimize their food production. The worm also passes its nitrogen-containing waste material to the microbes, allowing them ready access to the nitrogen necessary for building cellular components.

Similar relationships exist between sulfur-breathing microbes and the giant clams and mussels that thrive in the vent communities. These partnerships are very much like the ones that exist in the reef communities, except that the whole ecosystem is sustained by the energy derived from hydrogen sulfide, a compound that most living creatures would find highly toxic!

An ecologic disaster
The *Exxon Valdez* ran aground in the pristine waters of Prince William Sound, spilling millions of gallons of oil into the bay and devastating the environment.

Nature's Balancing Act

In March 1989, the *Exxon Valdez* supertanker ran aground on Bligh Reef in Prince William Sound, spilling over 11 million gallons of crude oil into the pristine waters of Alaska. The oil contaminated over 1,000 miles of shoreline. It was the largest oil spill in North American history.

Volunteers and professionals alike rushed in to save the wildlife and the shoreline. They tried to wipe the Sound clean of the contaminating oil, holding Exxon accountable both financially and ethically. But on the surface, most of their efforts seemed in vain. The toxic oil ultimately led to the death of more than 250,000 waterfowl, 2,000 sea otters, 300 harbor seals, and 250 bald eagles. It dramatically altered the fishing grounds that provided financial security for the native populations.

Within a few weeks, however, an amazing story began to unfold. Divers went down into the frigid water of the Sound to examine the damaged hold and ready the ship for towing. Much to their amazement, they discovered that the water that had flowed into Hold 3C, the damaged tank that had poured the toxic petrochemicals into the water, was alive with marine life. There were jellyfish, crustaceans, and mollusks. Herring and salmon swam about amid the microbial mats, marine worms, and algae now populating the cavernous ship. The hold that had contained nothing but oil was now home to a thriving ecosystem!

The death of birds and animals caught the attention of the world.

What had happened is a stunning example of earth's ability to heal itself, even in the face of a massive injury like the oil spill. Naturally occurring microbes, responding to the unexpected banquet of hydrocarbons, grew more rapidly than normal. Their usually small population increased dramatically, and they soon began to "eat" the oil, changing the toxic substances into cell material that would ultimately enter the food web in a benign form. They, in essence, did something that all of our frantic efforts were unable to accomplish. The microbes formed the foundation for restoring nature's balance in the most difficult of environments.

Spraying the oily shores with nitrogen to encourage the growth of petroleum-eating microbes turned out to be more efficient than the use of solvents in cleaning up the oily mess.

We learned many important lessons from the disaster. Perhaps chief among them was the natural resiliency that biologic systems have to restore their surroundings. Although arguably there are still residual effects from the spill, microbes changed an oily nightmare back into a thriving ecosystem.

We also learned that our most useful intervention was not washing oil from the coats and feathers of the wildlife, but rather supplementing the microbes' petroleum feast with critical limiting nutrients like nitrogen. The petroleum-eating microbes could only transform the oily black mess into their cell material as long as they had all the other elements necessary to complete their complex organic molecules. By adding enough microbial "fertilizer" to match their crude oil diet, the rate at which the oil was removed was increased by up to fivefold.

Because of the combined efforts of humans and microbes, restoration was shortened to 2 to 3 years from the anticipated 10 to 20 years. This strategy has become the one of choice for oil spills today. Visitors to Prince William Sound are often surprised to see the surroundings returned to their original splendor. While the long-term effects on this ecosystem's balance are still being sorted out, it's clear that life has not only survived but is thriving in this Alaskan wilderness.

Such experiences should not, however, make us complacent about the impact human activities have on our planet. As we burn fossil fuels, convert grasslands and forests to concrete and asphalt, and tie up resources in ways that slow their recycling, we should bear in mind that we are uncertain as to how robust earth's healing powers really are.

One lesson remains clear, however, from our experiences in Prince William Sound—microbes are a valuable counterbalancing force for at least some of the consequences of our clumsy activities.

The Sum of the Parts

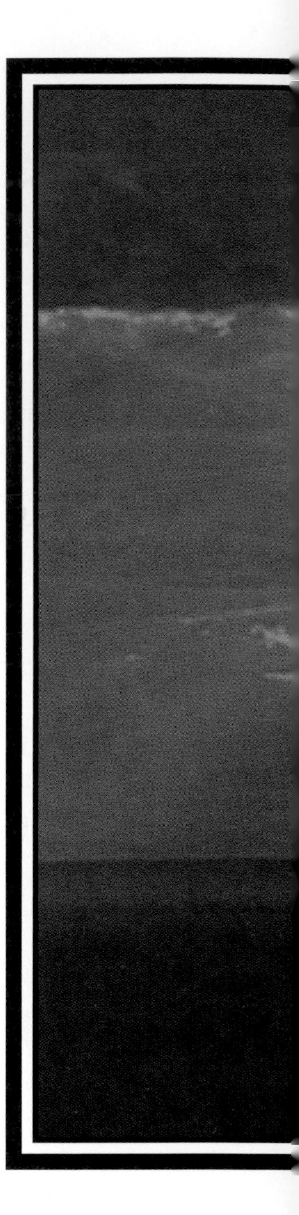

As remarkable as each individual ecosystem is, interconnected by microbes and composed of living arrangements that depend on massive exchanges of food and energy, the sum of the parts is truly something to behold. Each of earth's ecosystems, from the cold tundra and mountaintops through the temperate and tropical forests, grasslands and deserts, swamps and coastal waters, to the deep ocean, is ultimately connected to the rest in a balanced set of actions and reactions that define the conditions for life on earth.

Any physical environment selects for certain kinds of living creatures, but the interactions of living creatures in complex communities affect the physical environment. We can thus view the earth as a continuously evolving system in which both physical and biologic forces interact to create change. Somehow, this overwhelming complexity all balances out to create a dynamic equilibrium for the planet as a whole.

Understanding how this balance is achieved and maintained is critical for the future of humanity. The earth's 4.5 billion years of geologic history clearly demonstrate that these forces can create massive changes. It seems equally clear that we, in fact, are now influencing the amount of variation on a global scale through our own activities—an influence that may not, ultimately, be beneficial to our own self-interests.

The experiences from Biosphere 2 emphasized dramatically the degree of our ignorance about the way the earth maintains its dynamic equilibrium. The attempt to create a self-contained living environment for humans failed for many reasons, but it is notable that even though larger plant and animal life died off, microbial life continued to evolve and thrive.

Scientists have barely begun to explore the connection between the physical characteristics of our planet and the activities of the microbial world. Their early discoveries have been remarkable and offer some early lessons to consider. In thinking about how our planet works, it should be clear that microbes play a significant role, and that we need them more than they need us. Although they are the tiniest of life forms, collectively they can bring about large changes, including the warming and cooling of our planet.

If we are to understand how our world works, we must first understand theirs.

SECTION TWO

TREE OF LIFE

In all things of Nature, there is something of the marvelous.

—Aristotle, *Parts of Animals*

We all share a curiosity about our family tree. Many of us construct elaborate histories of our families, going back generation after generation to see where we came from and with whom we might share a common ancestor. We base our family trees on birth records, but we can also see certain physical characteristics that link us to our ancestors. We might have, for example, our aunt's dark, curly hair and brown eyes or our great grandfather's long, straight nose.

What if we had a way of tracing our family tree all the way back to the beginning of life on earth? How far would we have to go back? Who—or what—would we find there? And what could we discover about life?

We can easily extend our quest beyond the immediate members of our family. We can take a good look at our next-door neighbors and know that we are in fact related. We all share certain physical characteristics. We stand upright, we have opposable thumbs, and are mostly hairless. Similarly, we can look at the family cats and know that we are not as closely related to them as to our neighbors.

Observing basic similarities and differences, in fact, is the way we first began categorizing things around us. Plants were quite different from animals, fish were noticeably different from dogs, and beetles were more similar to grasshoppers than to birds. Very early in our written history, people could and did arrange things into categories based on their physical characteristics. Over 2,000 years ago, Aristotle arranged living things into two major groups—one for plants and one for animals—ordering living things from the most complex Celestial Beings down to what he considered to be the simplest and lowliest creatures.

From Aristotle's time until well into the 18th century, some people believed that all species were placed on earth "in the beginning"—some few thousand years ago—in their modern form, and that they had simply reproduced over the centuries. It occurred to no one that some species might have existed and become extinct. The notion that you might trace family trees of species back through time to a common ancestor for all living things was totally unthinkable.

We have always contemplated and told stories about our origins. We based our earliest stories on many things—sometimes on religious philosophies and sometimes on observations of the world around us. As scientists improved their powers of observation and discovered the meaning of fossils, however, our older views of the tree of life were shaken, and a new tree emerged. Now scientists are able to look into life's information set—the DNA—and are using it to build yet another tree of life, changing our views once again of how living creatures are related. With our increasing powers of observation, we can see that even our branch of the tree contains the genetic footprints of our earliest ancestors—the primitive microbes that gave rise to all living things today. The tools of modern genetics are revealing evidence that links us all to our common past and are helping us look to our future.

Shaking the Tree

The old, static view of life changed radically in the 18th and early 19th centuries as the science of geology came of age and paleontologists learned the nature and meaning of fossils. The earth proved to be far older and far more changeable than we had believed. We learned that fossils were the petrified remains of previously living things, some of which had no obvious living descendant. As scientists began combining the geologic records with the fossil records, a major paradigm shift emerged: both the earth and the life forms on it had changed prodigiously over time, and various life forms, both living and extinct, could be arranged along the branches of a single ancestral tree. Scientists were taking the first steps in understanding and building a modern tree of life.

In the mid-19th century, Charles Darwin and Alfred Russel Wallace articulated the theory of evolution and provided us with a mechanism for biologic change over time—natural selection. Darwin and Wallace were biologists, avid collectors, and keen observers of nature. They independently came to the same conclusions, and together they forever changed the way we viewed the living world. Their ideas may be summarized as follows:

- Life had a common origin, with new forms of life branching off from older forms.
- There is random variation among individuals within a population; such differences continually arise by chance.
- The pressure of a constantly changing environment in which individuals must compete for survival results in selection of favorable traits. Individuals possessing traits that fit well with the environment survive and pass those traits along to their offspring, while individuals with traits that do not fit perish.
- While each adaptation may be small, cumulative selection of favorable traits leads over time to increasingly different forms of life and eventually to new species.

By the 1930s, the scientists studying the anatomy of living creatures, the geology of the earth, the fossil records, and genetics had firmly established the tenets of Darwin's and Wallace's theory. In the 1950s, scientists began studying the chemistry of life at the molecular level. Their new approach to evolution has confirmed the basic precepts of evolution while dramatically extending our understanding into new areas that were unseen and unknown throughout most of our recorded history—the invisible world of microbes.

Growing understanding of earth's prodigious age and changeability, and of the significance of fossils, shook earlier views. New representations needed to show how living creatures were all related.

From Organizational Chart to Evolution

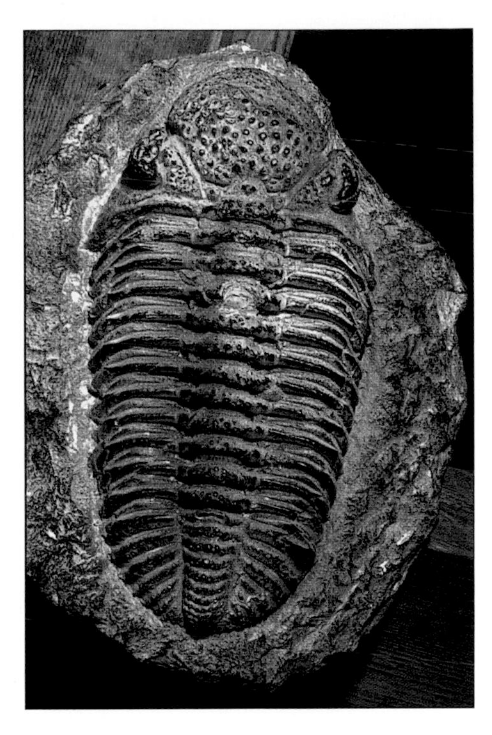

The trilobites, which lived 500 million years ago, are abundant in the fossil record because of their hard shell. More ancient creatures were all made of soft tissues that generally were not preserved in rocks, and so few traces of their existence remain.

Darwin's theory of evolution ignited an explosion in the field of paleontology. Prior to Darwin, Aristotle and others drew trees that were organizational, showing the relationship among currently living things. After Darwin, the tree represented evolutionary descent. Fossils came to make sense as milestones on our evolutionary journey.

But the fossil record posed a problem. The record is a history written in shells and bones. It goes back 600 to 700 million years to the earliest trilobites. But the trilobite was a distinctly complex creature. Since in evolution, the simple generally precedes the complex, a simpler creature must have preceded the trilobite. The earth itself was known to be some 4.5 billion years old, so there was an enormous stretch of time between earth's formation and these earliest fossils.

The microbial world was missing from the fossil record entirely. Bacteria, fungi, and algae were given branches on the tree, but there was no way to evaluate their evolutionary history and how they might relate to each other and everything else. And bacteria, fungi, and algae looked so much alike that they mostly defied classification by the traditional methods based on physical characteristics. Until the 1960s we had no way to explore the role of microbes on a tree of life.

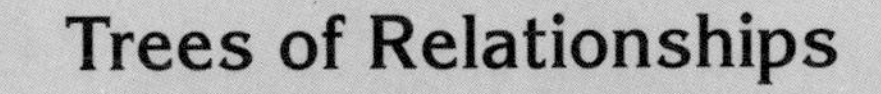

Trees of Relationships

A Darwinian tree of life portrays relatedness among different groups based on the assumption that all life had a common beginning. Each branch represents an evolutionary pathway, moving in a series of small, cumulative steps from earlier to later forms of life. Groups alive today are represented at the ends of the branches. The length of a branch can be equated roughly with both relatedness and time; the shorter the branches connecting two groups, the more closely related, and thus the more similar they are. At the point where the branches connecting two groups meet, they would have been indistinguishable from one another.

Consider tigers and lions and wolves. A tiger and a lion resemble each other more than either does a wolf. Based on their anatomies and the fossil record, we know they arose from a common ancestor. So we place them on a tree showing all three arising from a common branch point. The tiger and lion are more similar and thus separated from their common ancestor later than they did from the wolf, so we place them on branches arising from another, later, common point.

The 3.5-Billion-Year-Old Microbe

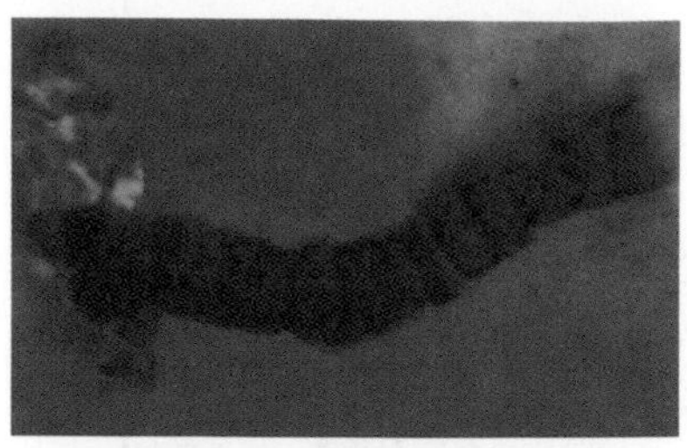

Identity: Fossil
Residence: Early Archaean apex basalt chert in Western Australia
Favorite pastime: Avoiding the killing rays of ultraviolet light
Activities: Probably an early ancestor of The Oxygen Generators, these microbes existed when broad, shallow seaways dominated the landscape; we suspect they may already have mastered some elements of photosynthesis.

Adding 3 Billion Years to the Tree

A breakthrough came in 1965 when scientists found fossilized remnants of microscopic single-celled organisms in rock formations that dated back 2 billion years. Microbial fossil hunters began to comb the oldest known rock formations, ranging from Greenland to Africa. The more they looked, the more fossil evidence of early microbial life they found.

J. William Schopf, a noted paleobiologist who works at UCLA, holds the record for the oldest fossil. He and his colleagues found evidence of microbial life in rock formations of the Apex chert in western Australia. The rocks are almost *3.5 billion years old*! The age was surprise enough, but Schopf found that the microbial life forms were far from simple even then. He and his colleagues identified at least 11 different types of microbes in their rocks, 6 of which look very much like a group of complex microbes found today called cyanobacteria.

Given what we know about the slow pace of evolutionary change, it would seem impossible that both the number of diverse cell types and the sophisticated cell structures in Schopf's fossils evolved over a short period.

The way life was . . . and is

Modern communities of microbes, growing in mats and called stromatolites, look very similar to ancient stromatolites that became fossilized billions of years ago. These bizarre microbial pillars can be seen in select locations around the world.

Scientists interpreted this information to mean that primitive life forms must have been present well before the 3.5-billion-year-old fossils were alive.

Finding microbial fossils that old had significant implications. First, the microbes must have been able to thrive in the most extreme conditions. The geologic records paint a picture of a primitive earth that was much hotter than today, with temperatures hovering near the boiling point of water. The atmosphere was devoid of oxygen, and there were a variety of gases present that we would find quite toxic—methane, carbon dioxide, nitrogen, and ammonia. Furthermore, since the first life on earth was microbial, and in fact exclusively microbial for the first 3 billion years or more, it followed that all of life's diversity—all the plants, animals, insects, and microbes—evolved from these primitive single-celled creatures and were relatively new on the scene.

This realization—that all members of the visible world alive today, including at least 280,000 animal species, 250,000 plant species, and 750,000 insect species, evolved from tiny, single-celled creatures—is astonishing to many. However, underlying all of life's diversity is the feature of life's biologic unity. Life's biologic unity brings together the invisible and visible world in a universal tree of life.

Cyanobacterium

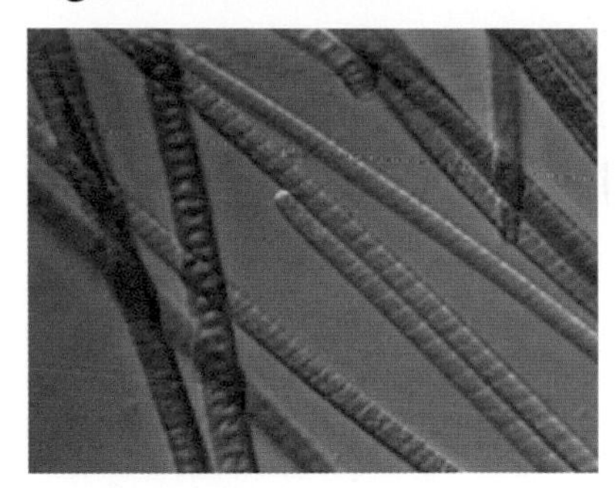

Identity: Bacterium
Residence: Just about anywhere there is sunlight
Favorite pastime: Making oxygen
Activities: Members of the Oxygen Generators, *Cyanobacterium*'s early relatives were responsible for changing earth's atmosphere from oxygen free to oxygen rich.

Exactly How Long Ago Was 3.5 Billion Years?

"It is difficult for us to appreciate exactly how far back life has existed on earth. One way to look at it is to imagine that all 4.5 billion years of geologic time is compressed into one 24-hour day. During the first second past midnight, earth forms. The origin of life would have occurred around 4:00 a.m. The oldest fossils known would have been deposited at 5:30 in the morning, just when the sun is coming up. The age of microscopic life starts then and continues all the way past noon, past 3:00, past 6:00, and in fact, it isn't until 9:00 at night that larger organisms appear. All of that evolution that you normally learn about in textbooks, from Trilobites to fish to amphibians to reptiles to birds and mammals, all takes place in the last 3 hours of the day. Humans don't appear until just a scant few seconds before midnight."

—J. William Schopf

Life's Diversity

One of the principal driving forces in life is the search for energy—food. In nature, of course, there are lots of different things to eat. However, food supplies are inherently limited. Creatures that are better adapted to exploiting a given food source—a niche—will get the largest share. Other kinds of creatures with small differences among individuals may find themselves with a competitive advantage in a slightly different niche, where they can then consume a larger share. If the second niche is sufficiently abundant and geographically separated from the first, the two groups may evolve in different directions, eventually becoming different species.

Exploiting new niches can lead to remarkable changes in a group of related creatures. Consider the Galapagos Islands. Darwin discovered at least 14 different species of finches there that appeared to have descended from a common finch ancestor. Each species arose because of a competitive advantage gained by adapting to a different food source. The most visible consequence of such adaptation is a different beak design to optimize food gathering for each species. The finches that eat insect grubs have long, pointed beaks for probing in the bark of trees; the finches that eat seeds have short powerful beaks for cracking the seeds open, and so forth. The biologic consequence is the evolution of different species.

Microbes, like finches, have evolved to gain competitive advantage in different niches. However, microbial evolution is far more difficult to chronicle. Unlike creatures of the macro world, microbes come in a limited variety of shapes and sizes and they seldom leave fossil remains. So physical features don't help much. Scientists have turned to another, quite different, method of assessing relatedness—the accumulation of differences in genes.

Darwin's finches

It seems likely that sometime in the past a few finches migrated from the mainland to an island in the Galapagos. The different types of finch now found on the islands, with beaks adapted to the available food, are descendants of that original pioneer species.

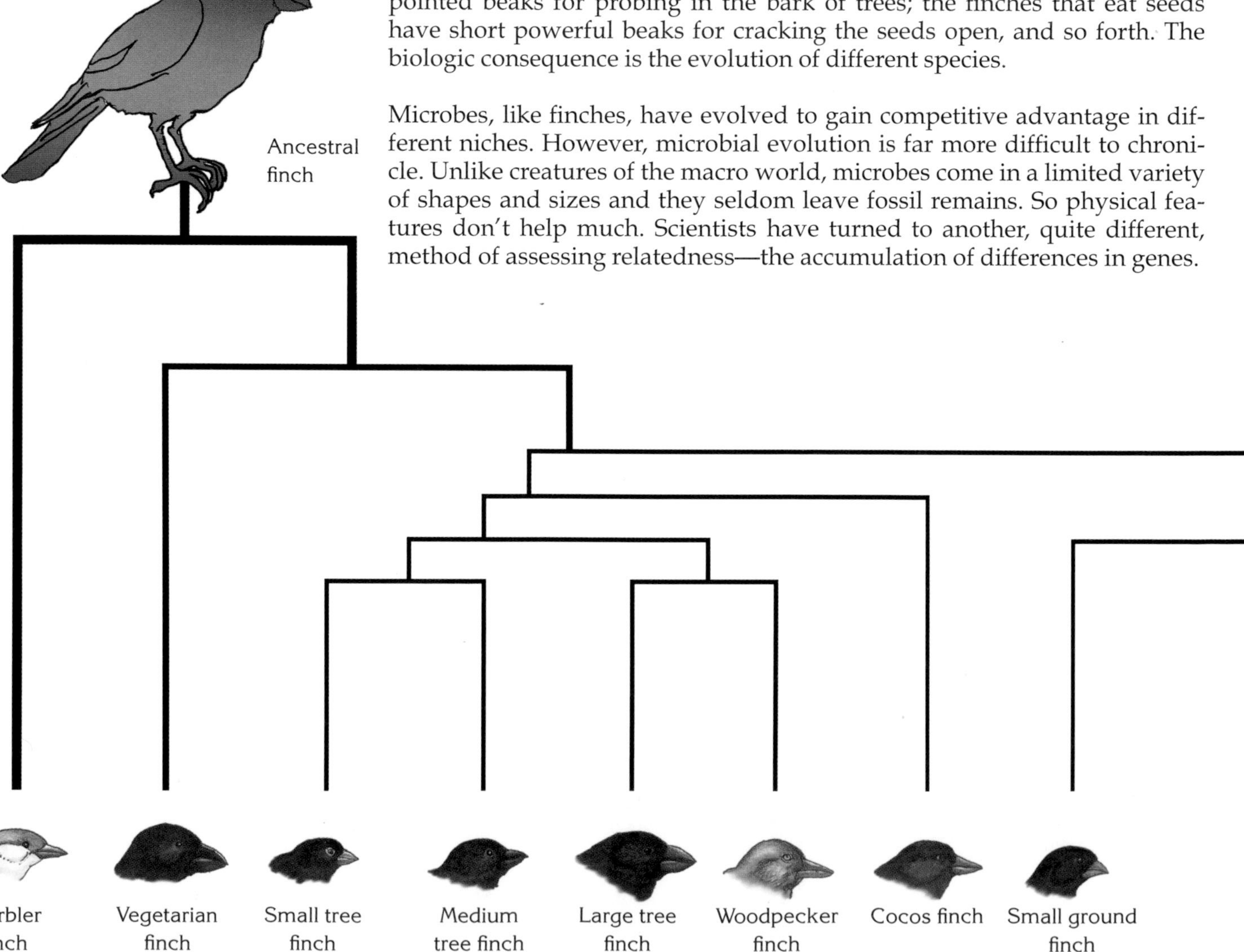

Diversity Springs from Unity

Although the appearance of all the various life forms on earth, and even some of their internal chemistry, may be different, their central organization is remarkably similar. The key feature that all living cells have in common is the separation of information and machinery into two distinct molecular realms. In one realm, the "instructions" for creating and maintaining the whole organism are stored and passed from generation to generation. In the other, these instructions are translated into the proteins—the "machinery"—that perform the tasks of life: generating and using energy, building and maintaining the cell, and reproducing. All life, from finches to microbes, shares this basic design.

The instructions are DNA. DNA is made up of long chains of four different chemical units called nucleotides. Humans have about 3 billion of these chemical units forming their instructions, while finches have fewer, and microbes fewer yet. DNA nonetheless operates in exactly the same manner in each. The long chains of nucleotides are organized into small sections, about 1,000 units long, called genes. A gene contains the instructions for making a protein—life's machinery. Just as the instructions for playing the notes in a particular passage of music are contained in the order of notes in a musical score, the instructions for making a protein are based on the order of the nucleotides in the gene.

The opportunity for diversity to evolve lies in changes in a gene's instructions. The nucleotide units that make up each gene are subject to random changes called mutations. Just as changing a note in a musical score will change the sound of the musical passage when it's played, a random change in one or more of the nucleotides in a gene will change the set of instructions for making a protein. The first produces an altered melody, the second produces an altered protein.

Individual protein molecules are robotlike performers of simple tasks. A living creature, even one as small as a bacterium, is the result of the coordinated interaction of thousands of proteins, each responsible for a different job. Because of the interconnectedness of their work, changing one protein may change the work of others. This means that a single mutation in a gene can lead to a very large change in living creatures.

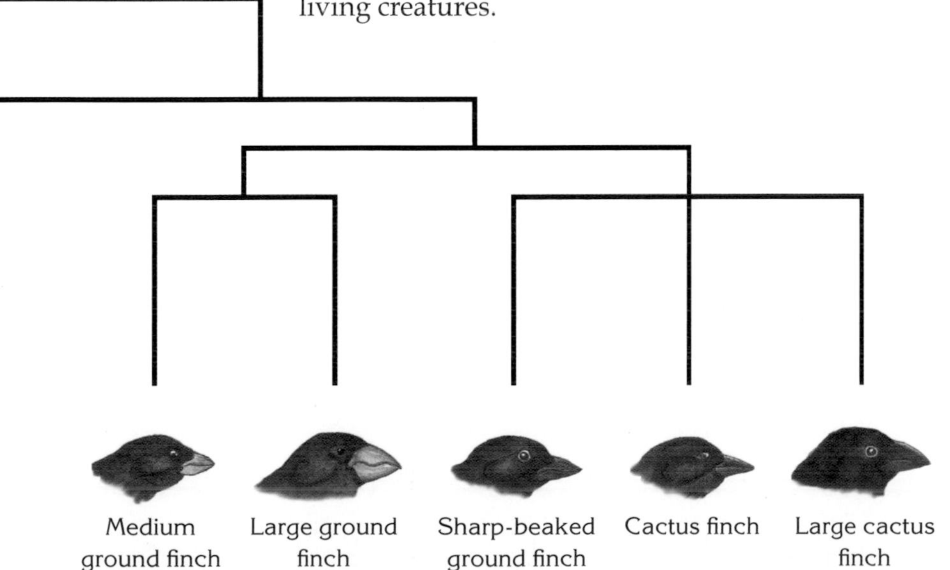

Life's unity

Every cell of every living creature carries within it two basic classes of molecules: information molecules (DNA), which carry life's ideas, and machinery molecules (proteins), which do life's work.

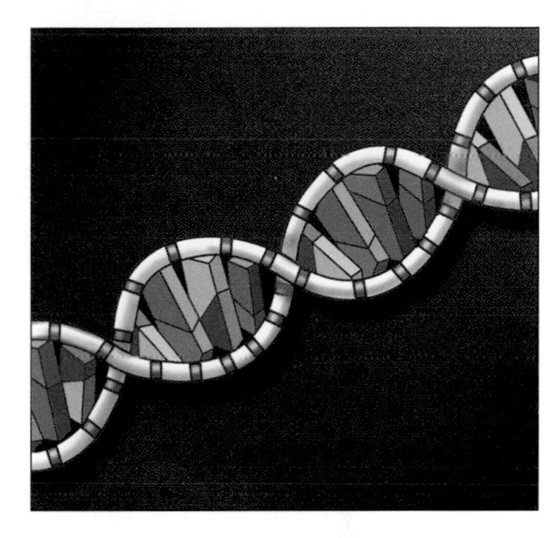

Information molecule

DNA's sequence of four nucleotides is a set of instructions for making life's machinery.

Machinery molecule

Proteins—thousands of different kinds—are life's infrastructure, machinery, and moving parts. Among their functions is the reading and replication of DNA.

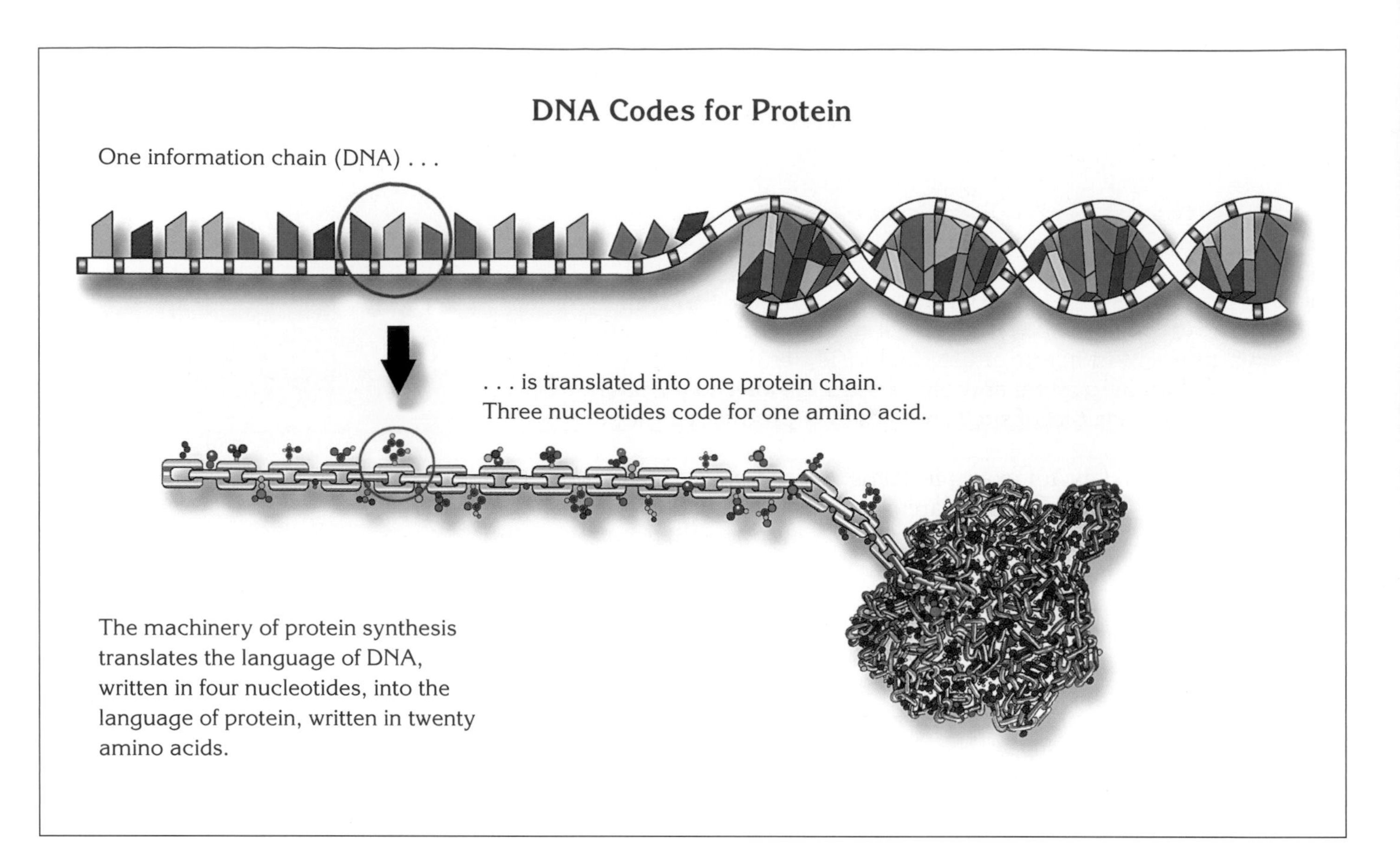
DNA Codes for Protein
One information chain (DNA) . . .
. . . is translated into one protein chain.
Three nucleotides code for one amino acid.
The machinery of protein synthesis
translates the language of DNA,
written in four nucleotides, into the
language of protein, written in twenty
amino acids.

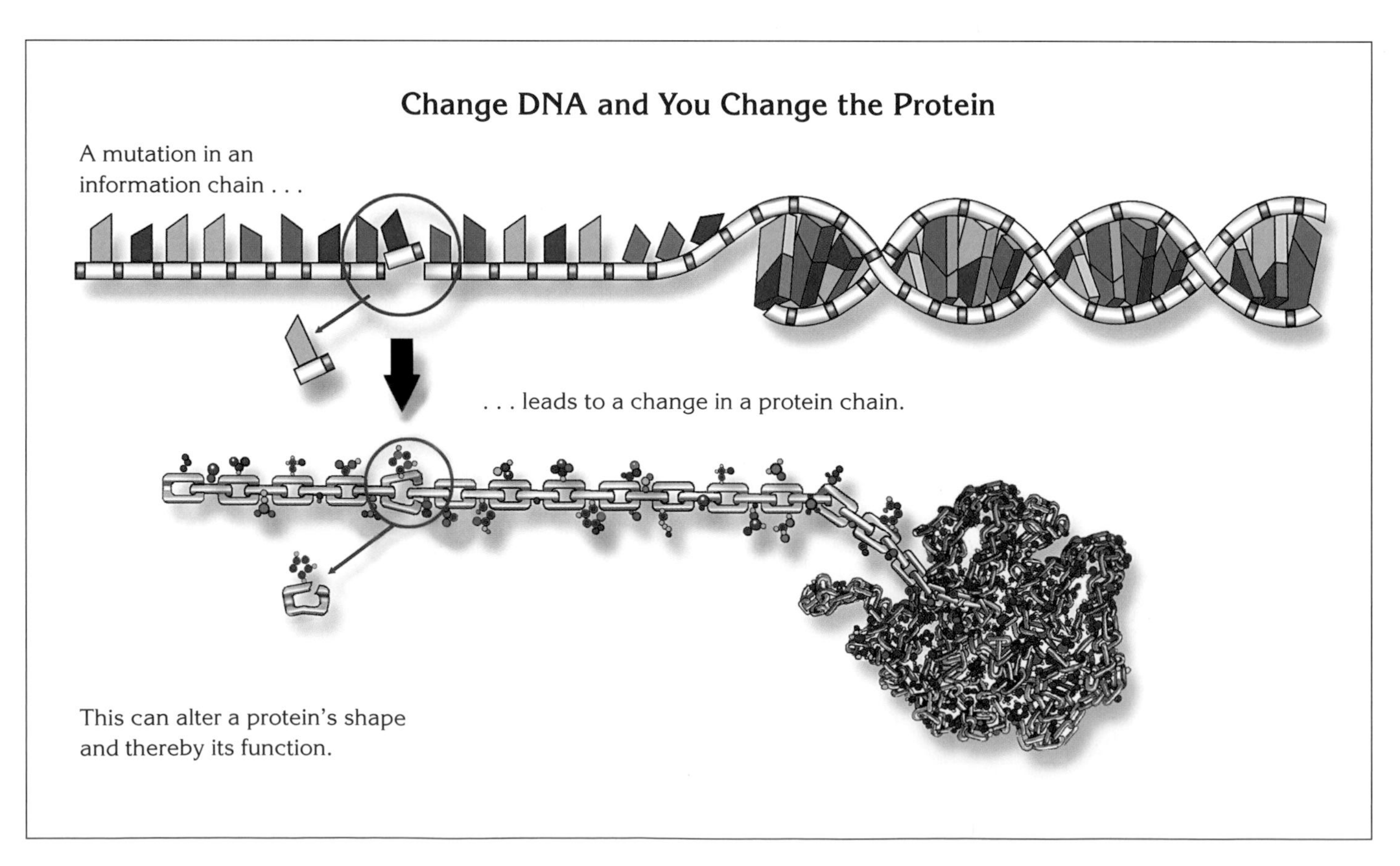
Change DNA and You Change the Protein
A mutation in an
information chain . . .
. . . leads to a change in a protein chain.
This can alter a protein's shape
and thereby its function.

From Genes to Proteins to Beaks

Most mutational changes damage a protein's function. Once in a great while, however, the protein will work better. Such an improvement is at the root of natural selection. An individual with the improved protein may thrive better than its parents and pass its traits along to its progeny. Over time, the cumulative effect of such changes may produce a new species.

Such small changes may, for example, translate into physical characteristics like beak design. Some proteins are structural, and others control the amount of structural material produced. The latter act like switches; the longer they are on, the more structural material will be produced.

Finch speciation is the result of multiple changes in proteins that guide the development of different body parts.

A random change in the DNA coding for a switch may change the length of time that it is on. As a result, the amount of structural protein laid down during the development of the beak changes its final shape.

Imagine that a population of thin-beaked, insect-eating finches in the Galapagos Islands find themselves in a new environment: perhaps the climate changed or they migrated to a new island. Suddenly the most abundant source of food is hard-shelled seeds. Some among their number will have a sufficient difference in their genetic makeup (brought about through small changes in their DNA) to possess larger and tougher beaks than others, and consequently will be more successful in cracking seeds. These finches will thrive and produce more offspring with strong beaks. In time, they will become the majority.

If this better-adapted beak were, by chance, associated with the production of more musculature (a protein) in the neck, an even more potent seed-breaker would have emerged. Small changes in the protein enzymes in the bird's intestine, making the core material of seeds more digestible, would further a bird's and its offspring's success. Eventually, small changes in the finches' DNA would thus be translated into small changes in the finches' characteristics, and new species would emerge.

Switches control size and shape

Imagine a frozen yogurt machine with four pipes, each with its own control switch.

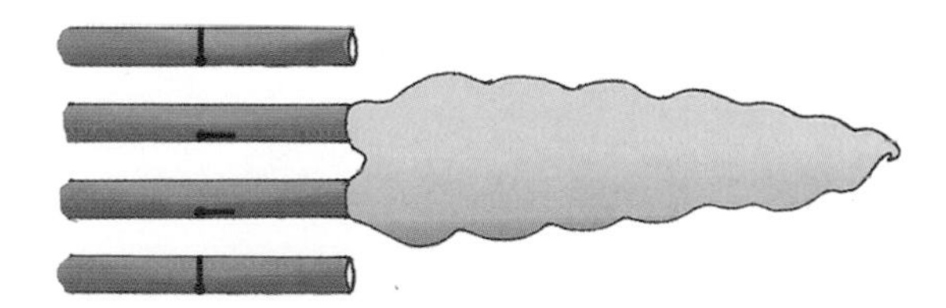

The two center pipes, with their switches fully opened, extrude a long thin shape.

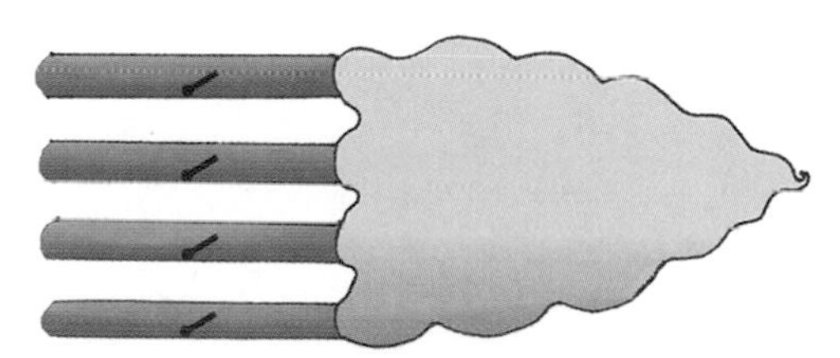

Four pipes, with their switches partially opened, create a shorter, thicker shape.

In reality, proteins, acting as genetic switches, control the amount and location of beak material and thereby its size and shape.

Microbial Diversity and the Volkswagen Syndrome

As already noted, physical characteristics like the finches' beaks are not so obvious among microbes. In fact, if we categorized microbes only by their external features, we would have a very small number of groups compared with the multitude of groups within the visible world. We would be wrong. We would be faced with what a famous paleobiologist has dubbed the Volkswagen Syndrome.

The Volkswagen Syndrome is what happens when scientists make assumptions about similarity based only on appearance. The Volkswagen Beetle of the 1940s, 1950s, and 1960s had pretty much the same body design year after year. Yet year after year, internal engineering changes were made so that the newest model was always different. The new Beetle might have more horsepower or a fancier carburetor. The insides were different even though the outside looked almost exactly the same. Just like the Volkswagen Beetle, microbial diversity, the number of different groups, is based less on body design and more on what's "under the hood."

The engineers at Volkswagen designed the Beetle to fit a very specific niche. It was to be an inexpensive car for the working man. In order to continue to fill that niche, the engineers made the outside of the little car almost the same year after year. It worked well, and changing it would have imposed unnecessary costs. Rather, the designers chose to focus their energy on improving the performance of the car by changing its inner workings. Microbial evolution may have followed the same general tactic, improving the inner workings while leaving the outside well enough alone.

Don't be fooled by the shape

Over their billions of years of existence, microbes have retained their outer shell while working on improvements within. Although these bacteria look very similar, one lives in the human gut, making vitamins; one lives at the bottom of the ocean, "breathing" sulfur; one lives in soil and converts sunlight to energy; and so on.

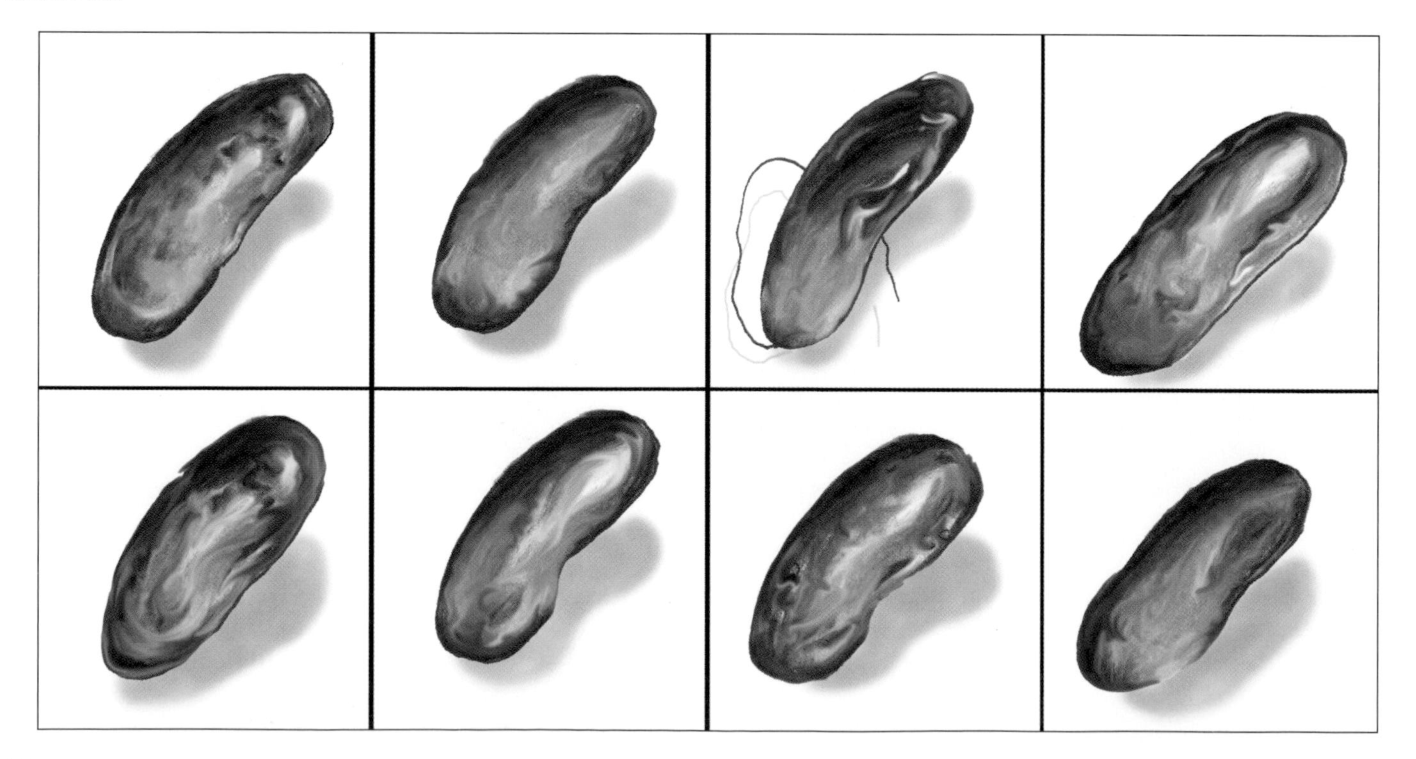

Unlike the Volkswagen engineers, microbes have had several billion years to experiment—their inner workings evolving to exploit almost every conceivable energy source as food, thereby adapting to virtually any and every niche. We would expect the microbial portion of the tree of life to be rich and complex as a consequence. It is. Scientists now believe that thousands of different microbes have adapted to live in every environment, and that there are far more different kinds of microbes than all other life forms combined.

Some microbes survive successfully without light or oxygen, using sulfur as food. Others can literally get energy from a stone. All of these microbial food-gathering strategies, however, came about not through intentional engineering, like the Beetle, but through the same processes that gave rise to the differences in finch beaks. The changes came about because of changes in the DNA that gave rise to changes in the microbes' proteins.

The Volkswagen Syndrome

Even though the outer shells of the Volkswagen Beetles looked remarkably alike, their internal features were quite different. They were continuously changed over their 30 years of production to improve performance. So it is for microbes.

	1940s	1970s
Engine	25 HP	50 HP
Brakes	Mechanical	Hydraulic
Shock absorbers	Single action hydraulic	McPherson strut
Fuel system	Carburetor	Fuel injection
Price	£85	£1,644

Using DNA To Build the Tree of Life

Differences determine relatedness

Three organisms have slightly different versions of the same gene, each represented below. Compare pairs of sequences and record the number of differences. Which pair of the three are (a) most closely related (the fewest differences) and (b) most distantly related (the most differences)? Comparisons like this help scientists build the genetic trees of life.

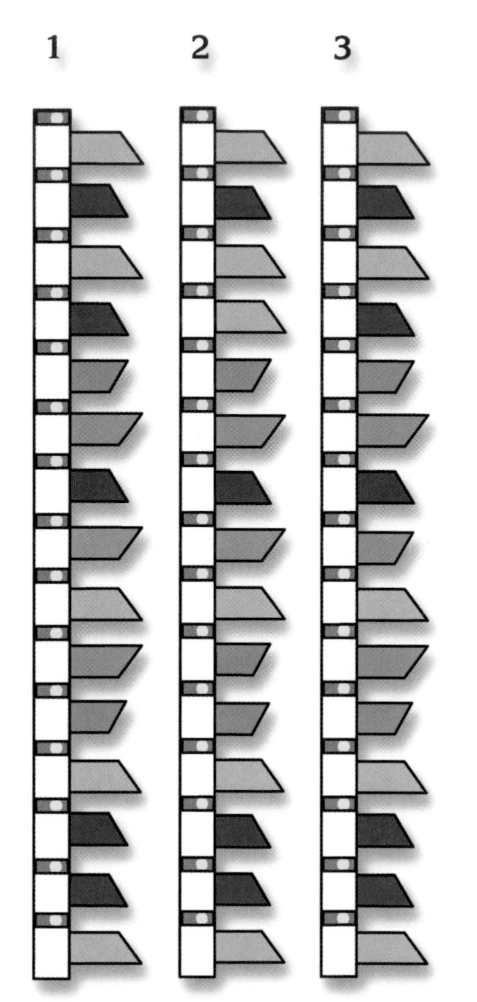

Number of differences

___ 1 vs 2
___ 1 vs 3
___ 2 vs 3

Answers: (a) 1 and 3; (b) 2 and 3.

Understanding the central duality of DNA and protein—and how small incremental changes in the former translated into small incremental changes in the latter—provided a new way of looking at relationships. Not only did this new approach gradually reveal a possible evolutionary path for the microbial world, but it provided evidence that all the visible life forms on earth shared a common ancestor.

The new approach relies on DNA. As we've seen, all living creatures use DNA as their own unique copies of life's instruction book, and DNA chains are subject to random changes (mutations) in their basic nucleotide units. DNA chains carrying the changes are passed to subsequent generations in the normal process of DNA replication. In the case of microbes, this is very straightforward, since most microbes reproduce by copying their DNA and then dividing in half. The two new cells each carry a complete set of instructions in their DNA, including any altered nucleotide sequences. Over time, such changes accumulate within the DNA of all living species.

Measuring the current state of this ongoing process in the life around us provides a simple, mathematical way to view relatedness. A gene in one species would be different from a gene performing the same function in a second species because of mutations that had accumulated after the two species branched apart from a common ancestor. By counting the number of differences in the gene from the two species, we can deduce how closely related they are. If there are few differences, the species are closely related; if there are many, the two species are more distantly related. By replicating this analysis for many species, we can construct trees of relatedness among species. In these trees the length of branches connecting any two species is proportional to the number of differences in a specific gene between those two species.

Scientists used this approach to examine evolutionary relationships among animals and plants. The family trees they constructed for living members of the visible world from such analyses are in remarkably close agreement with those based on comparing the anatomy of one species with that of another.

In other words, we had a reliable way of measuring relatedness that didn't rely on discovering a fossil record or on external appearances. This approach has become central to building a new hypothesis about how *all* life has evolved, including microbes. It has helped us understand and explain how life creates and supports the tremendous diversity of living creatures that share the earth with us.

A Mutation—Aiiee!

Contrary to the notion of movie screenwriters, a single change in DNA (sometimes called a mutation) does not result in an insect the size of a building. Nevertheless, over time, cumulative mutations and natural selection can result in creatures as dissimilar as mushrooms and mastodons.

The Waste Product That Caused the Greening of the Planet

With microbes, and probably everything else, the process of mutation and natural selection is a two-way street. As mutations quickly led to adaptations for different environments, the results of some of those adaptations began to modify the existing environment.

One of the most dramatic examples of environmental impact took place very early in the evolution of life. Through a long series of mutations and adaptations, microbes invented photosynthesis. Photosynthesis is the process that plants, algae, and some bacteria use today to take energy directly from the sun and, using carbon dioxide and water, convert it into cell material. When this sophisticated lifestyle first came into existence, it gave the newly photosynthetic microbes a powerful competitive advantage over their hydrogen- and sulfur-breathing relatives. Suddenly, there was an almost limitless supply of direct energy from the sun readily available to the photosynthesizing microbes. Their populations rapidly expanded, covering the surface of the planet and thriving wherever there was light and water.

The expansion of photosynthetic populations might not have been quite so bothersome to other, nonphotosynthetic microbes save for one factor. Photosynthesis generated a new waste product—oxygen! We don't usually view oxygen as a pollutant, but the microbes living on earth then were likely to have found it quite toxic.

The photosynthetic microbes thrived, and the concentration of their metabolic waste product, oxygen, gradually increased in the atmosphere to the present-day level of 20+%. Organisms used to the then prevailing level of 0.1% were forced to evolve mechanisms to detoxify oxygen or were driven into habitats where oxygen couldn't reach them. Or they simply died off.

Microbial generation of oxygen had another consequence: the formation of ozone, the important ingredient in our modern atmosphere that shields us from the harmful effects of ultraviolet radiation. Until ozone appeared, life could only exist where organisms were protected from the damaging rays of the sun—under rocks and under the surface of the water. With the buildup of the ozone layer, microbes could expand over the entire surface of the earth. And they did.

So microbes changed the face of the planet, and the conditions they created it for all the life forms that would follow. Shortly after the formation of the ozone layer, there was a massive evolutionary explosion, culminating in the origin of the first multicelled organisms.

Norman Pace and Carl Woese
in the laboratory at the University of
Illinois, Champaign-Urbana

Placing the Microbial Branches on the Tree

Carl Woese, an evolutionary biologist at the University of Illinois at Champaign-Urbana, realized if he wanted to probe evolution to its depths, he had to find a way to build a *universal* tree of life. The new tree would have to include all the single-celled organisms—the bacteria, the protists, the algae and fungi—along with the plants and animals. He reasoned that if he could find a gene common to all or most living creatures, including microbes, he'd have the key to the remotest past.

Finding the right gene was a tall order. He needed a gene whose product had a common function in all the cells in which it was found. He needed a gene that was very ancient and that was still present in all modern living cells. He needed a gene that had an ample number of nucleotides to see both changes and similarities in sequences. And, he needed a gene in which change would have occurred very slowly, so that common sequences would still be present, even across the broadest species differences and the longest stretches of time.

All known modern cells have a structure called a ribosome that translates DNA's information into proteins. Woese reasoned, therefore, that ribosomes were very likely to have been present in the first primitive cells—the cells from which all present life forms evolved—and that the genes for building them had been passed in the DNA information throughout evolutionary history.

A ribosome is built from two building blocks—proteins and ribosomal RNA. Although we normally think of RNA as an information intermediary for translating DNA into protein (messenger RNA), a certain kind of RNA (ribosomal RNA) is a part of the ribosome's structure. It's as though a composer rolled up a copy of her score and used it to construct a violin.

Ribosomal RNA is a direct copy of the gene that codes for it. That gene seemed to Woese to fulfill all the conditions he was seeking. If the ribosomal RNA gene was as ancient and critical to life as he suspected, Woese hypothesized that it would have changed only very gradually over time.

The ribosome

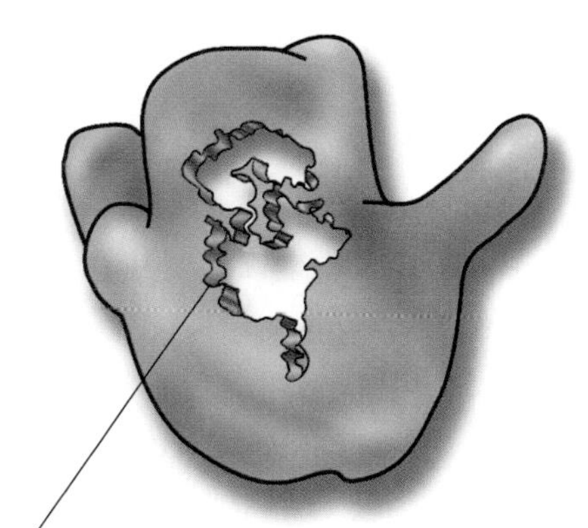

Ribosomes, found in every living cell, are made of protein . . .

. . . and RNA. The gene for the RNA chain provided the basis for Woese's analyses. It was just the right size to analyze and changed very slowly over time.

Woese began to test his hypothesis. He analyzed the ribosomal RNA gene from species after species, slowly deciphering their sequences of nucleotides.

He found that he'd hit the nail on the head. Parts of the sequence of the ribosomal RNA gene from every species he examined were exactly the same—the nucleotides were in exactly the same order. This suggested to Woese that all cells had a common ancestor. Other stretches of the ribosomal RNA gene were exactly the same for all the prokaryotic cells he examined, but those same stretches were markedly different in eukaryotic cells. This suggested to him that the prokaryotic and eukaryotic ancestor had diverged into two distinct groups. Within each of these groups, he could discern genetic evidence of further subdivision, hence extending the branches.

Building genetic trees

A fanciful look at a complex process

If the nucleotides were musical notes, Woese's findings would be like discovering that all the first living cells started out playing the same melody. Over the billions of years of evolution, each species modified the score a little, creating variations in the song and passing them along to their progeny.

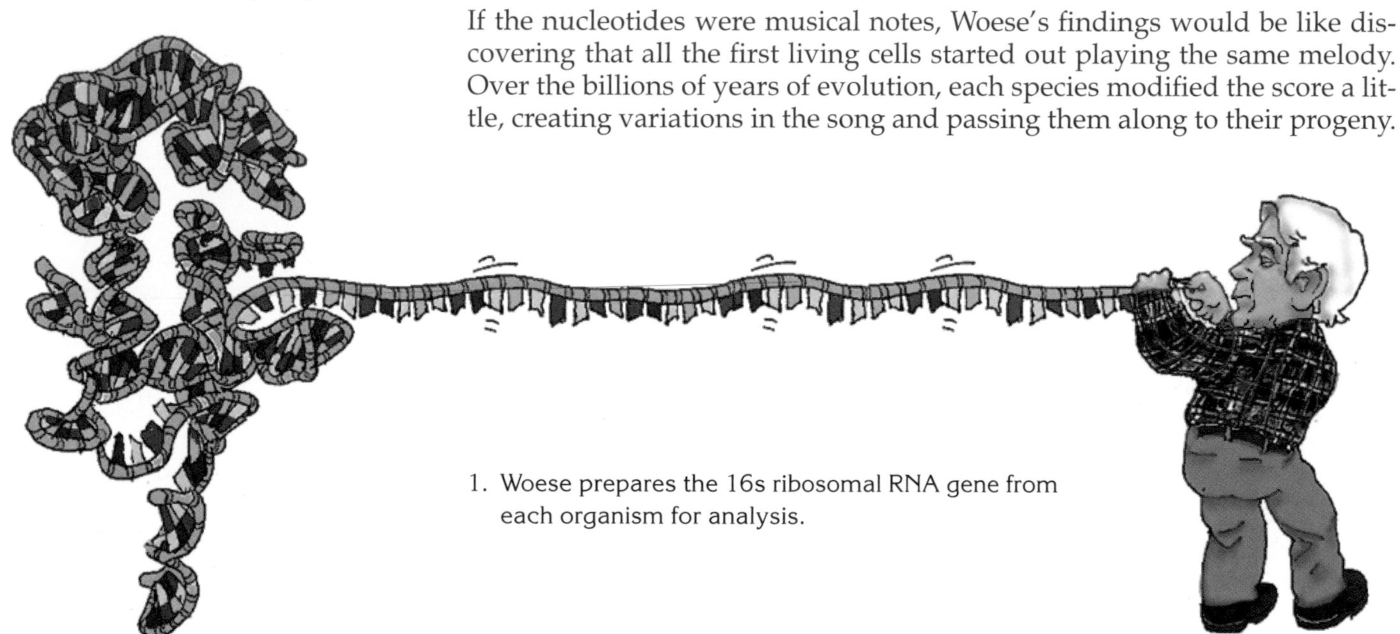

1. Woese prepares the 16s ribosomal RNA gene from each organism for analysis.

2. He compares the rRNA genes' nucleotide sequences from different kinds of organisms (8 shown here), identifying the stretches that are alike and those that are different.

During 3.6 billion years of evolution, there would have been plenty of time to create multitudes of variations, but we would still hear part of that original melody in every variation—in every species that exists today, from *Escherichia coli* to you and me.

Week after week, year after year of painstaking work, the variations in the sequences of nucleotides in the ribosomal RNA gene fell into place and Woese's new tree of life took shape. He searched primarily for the sections of the RNA molecule that indicated where the tree branched—where groups of organisms shared a signature sequence not shared by other groups.

Thus, this ancient gene for ribosomal RNA became the basis for a new, universal tree of life. As Woese continued to build this new tree, he saw only two major branches, called domains: the prokaryotes (organisms whose cells lack a membrane encircling their chromosomes) and the eukaryotes (organisms whose cells have a membrane around their chromosomes). But a particularly soul-satisfying surprise was waiting just around the bend.

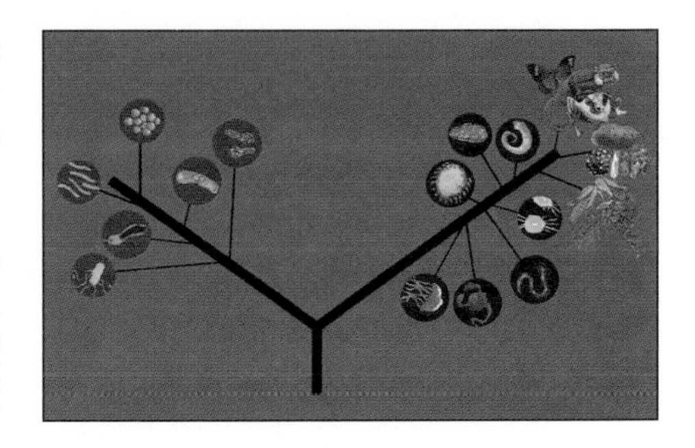

Scientists build genetic trees, in actuality, by using computer analyses that compare the rRNA gene's nucleotide sequence from one organism to those of many others. Such analyses yield trees with branches of varying lengths that diverge in different places.

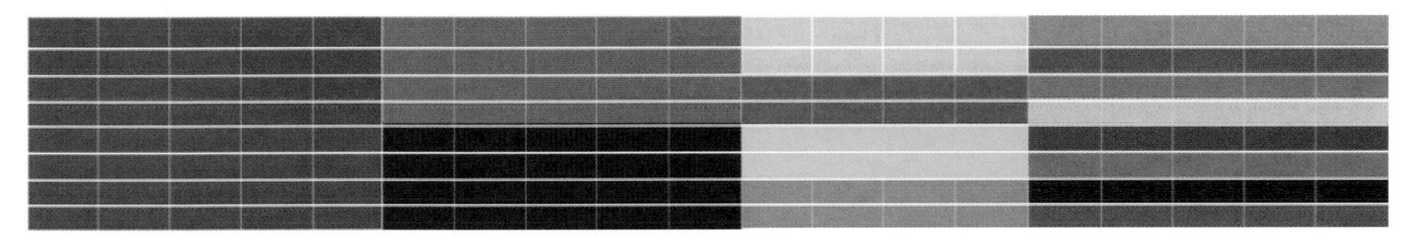

3. He arranges each organism's segments in order from the most common, universally shared color to the least commonly shared colors.

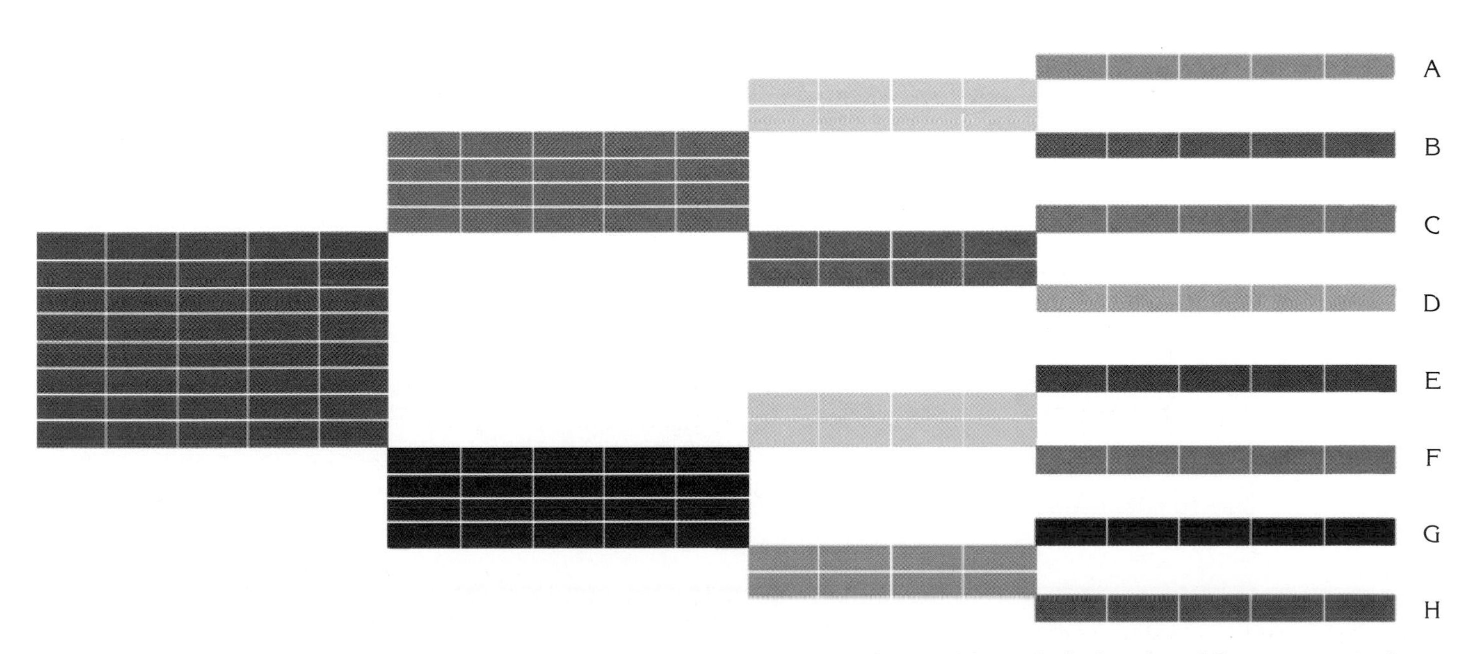

4. By spreading the segments apart, a tree emerges. In this example, scientists would conclude that A and B are more similar and therefore diverged more recently than did A and D.

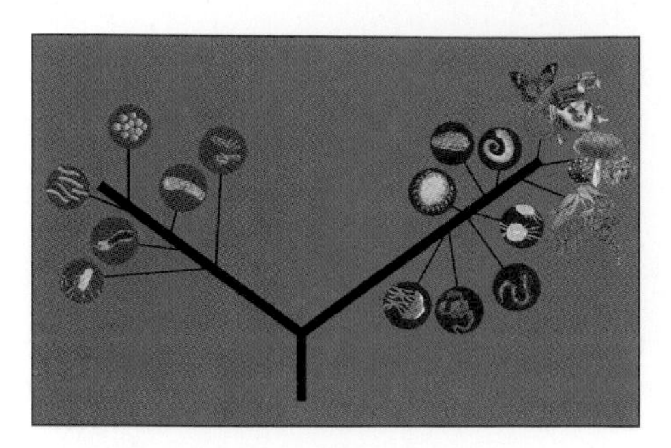

Early genetic trees divided the living world into two major branches—the prokaryotes and the eukaryotes.

Of Heat and Age: A New Branch Discovered

Midway through Woese's massive undertaking, a colleague, Ralph Wolfe, brought him a prokaryotic microbe that had some unusual characteristics. It belonged to a group of microbes named methanogens (methane generators). Wolfe's organism existed on a diet of carbon dioxide and produced methane as a by-product of its metabolism. More important, the single-celled creature liked to grow at very hot temperatures—temperatures that were close to the boiling point of water.

When Woese looked at this organism's ribosomal RNA, he was puzzled. The methanogen lacked the unique sequences found in all the other bacteria he had examined. The unique eukaryotic sequences were missing as well. This odd organism could not be fit into either of the two major branching domains. It was the first member of an entirely different domain of the tree—a group that Woese named the Archaea, from the Greek word meaning ancient.

Woese's choice of the name Archaea turned out to be propitious. Scientists, recognizing that the earth was extremely hot when life emerged almost 4 billion years ago, hypothesized that the more heat-loving an organism is, the closer it would be to the first life form on earth and therefore closer to the first split in the tree.

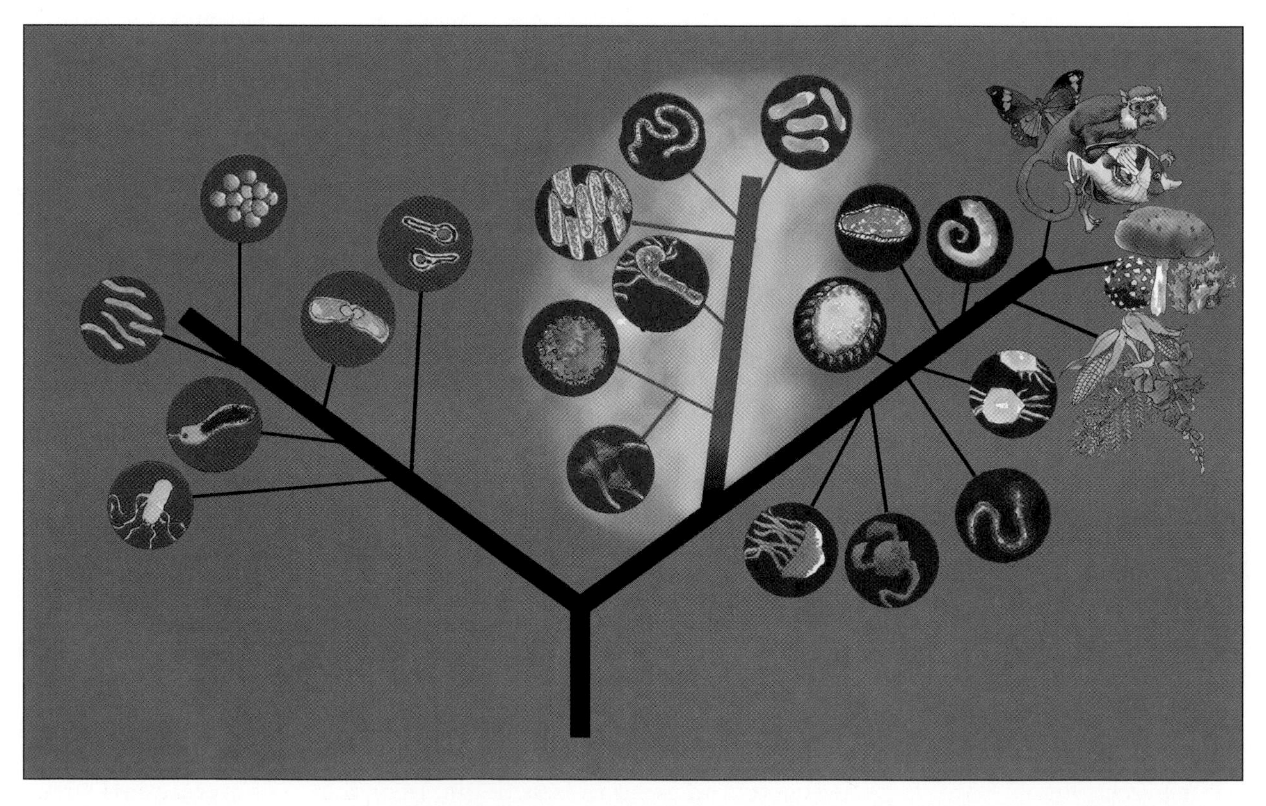

Midway through Woese's undertaking, he discovered a third major branching of the tree. The new tree had a branch for Bacteria, a branch for Eukarya, and a branch for the strange new microbes that lived in the hot vents—the Archaea.

Karl Stetter's long fascination with microbes that thrive in boiling pools and in steaming holes has brought him many times to the island of Vulcanos, off the coast of Sicily.

Karl Stetter is one of the scientists searching for the heat-loving microbes that he believes are most similar to earth's oldest microbes. He is also the most successful in the world at growing these high-temperature microbes, called thermophiles, in the laboratory. This is no simple feat, since thermophiles will only grow in conditions that simulate the ones found in their natural environment.

He calls his laboratory his "witch's kitchen." It's filled with pressure cooker-like vats that Stetter and his assistants maintain at very high temperatures and flood with atmospheres of hydrogen, nitrogen, and sulfur—rather like an indoor volcano. So far, he and his fellow scientists have created just the right environment for growing an extraordinary collection of microbes. Using his indoor volcanoes, Stetter grew the first microbe able to reproduce above the temperature of boiling water.

Stetter has traveled around the world in search of earth's hottest spots, bringing back samples to his laboratory. He is searching for the nearest living relative of the earth's oldest life forms. His search has taken him from the tops of volcanoes to the hydrothermal vents a mile below the surface of the ocean. He finds the microbes in boiling pools, in steaming holes in the earth, and along the shore in hot mud flats.

Hydrothermal vents offer a particularly interesting hunting ground. They occur at various spots along the Pacific and Atlantic sea floors, near the points where hot basalt and magma are very near the floor surface, causing it to slowly spread. As the crust of the earth opens, underwater volcanoes form, spewing out water heated to 350°C or more and laden with minerals.

Smokers

Hydrothermal vents support communities of organisms that depend on microbes to supply a usable form of energy—microbes believed to be similar to the earliest forms of life on earth.

The hotter the microbe's natural environment, the nearer it sits to the base of the tree, and hence the more similar it is to earth's first life forms.

These underwater volcanoes are called smokers, and they are surrounded by a complex and bizarre community of living forms, including tube worms 2 meters in length, giant clams, bright red shrimp, mussels—and, of course, microbes. Stetter recovered a hyperthermophilic microbe living on the side of one of these deep-sea vents that can thrive in temperatures up to 110°C, above the normal 100°C boiling point of water.

With each new microbe that Stetter discovered, Carl Woese found an important pattern. The hotter and hotter the temperature at which they could grow, the closer and closer the microbes were to the base of their domain's major branching point on his tree of life. Since many of these heat-loving microbes are Archaea, they may truly be ancient—in fact, very similar to the first living cells.

How a Microbe Got Its Name

Thermotoga maritima is one of the "hottest" microbes so far discovered on the bacterial branch of the tree of life. Scientists call microbes like this hyperthermophiles—"high heat loving." Karl Stetter was the first to get *T. maritima* to grow in the laboratory, and as is often the custom, he was the person who got to choose its name. When he looked at the microbe under a microscope, he saw a curious, loose-fitting socklike structure that covered either end of the rod-shaped bacterium. Since the organism was heat loving, his first idea was to name it "Thermosockus."

Stetter sent the microbe to Carl Woese so that he might determine its place on his tree of life. Woese found that it was as close to the major branching point as any bacterium that he had examined. When Stetter learned that he had potentially found a living organism that might be very similar to the most ancient bacteria, he decided that it must have a more dignified name. Being named after a sock was just not sufficiently distinguished for a microbe with such important relatives.

Stetter had found the microbe living in the volcanoes on the sea floor near Vulcanos, Italy, an area that was formerly a part of the Roman Empire. The socklike structure could just as easily look like a toga—the customary clothing worn only by the nobility in ancient Rome. On further reflection, he chose a new name for the microbe: *Thermotoga maritima*—a heat-loving, toga-wearing bacterium that lives in the sea (maritima).

And that, in many ways, seems a more appropriate name for a microbe perhaps most similar to one of earth's first living cells.

Going to Extremes

Think about jumping into a pond of boiling hot acid with a scuba diving tank filled with sulfur. Not very appealing to us, but microbes have evolved that thrive in these environments. We call them extremophiles, and the heat-lovers are just one example.

The enzymes they make, unlike our enzymes, are relatively resistant to heat and other noxious conditions. Called extremozymes, they are becoming increasingly valuable for commercial and forensic purposes.

Perhaps the most famous extremozyme is Taq polymerase, an enzyme that works at very high temperatures. Taq polymerase was the key to inventing an automated, high-temperature technique for making millions of copies of a single gene. This automated technique—called polymerase chain reaction (PCR)—is used extensively today in forensic medicine, in medical diagnosis (for example, HIV infection), and in screening for genetic traits such as susceptibility to certain kinds of cancer. Taq polymerase and PCR paved the way for Norman Pace, another noted evolutionary biologist, and his colleagues to make copies of ribosomal RNA genes directly from their samples, extending the tree of life well beyond the known microbial world.

Ranchers and farmers use extremozymes that work in highly acidic conditions as additives in animal feed. The extremozymes accompany the feed into the animal's stomach. Amid the stomach's acid, these extremozymes help break the feed down into smaller substances that can be more readily digested. The result is that ranchers and farmers can use less expensive grains and still provide the same level of nutrition.

Detergent manufacturers have a major interest in extremozymes that work under highly alkaline conditions. The common stains in clothes are usually either protein or lipid. Enzymes are very good at breaking down protein and lipid so that the stains can be removed, but most enzymes don't work well in the highly alkaline conditions created by detergent. Extremozymes that function in either hot or cold water and can withstand the alkaline conditions of the detergent are valuable tools for removing life's little messes.

"The distribution of life on the planet is very different than we would have thought it even a decade ago. We know that there's life shot throughout the crust of the planet. Life is in ice, life is in waters of boiling temperatures."

—Norman Pace

The Amplification of the Tree

While some scientists were off in search of the hottest microbes, others were exploring elsewhere in the microbial world, seeking to build an ever more complete picture of life within earth's biosphere. Most microbe hunters were troubled, however. Woese's analyses required large samples of the ribosomal RNA gene, derived from millions of microbes. This meant researchers had to be able to grow large quantities of microorganisms in the laboratory, starting with only a small number. The scientists knew that this simply wasn't possible with most of the microbes they could distinguish under the microscope.

Enter Norman Pace, a long-time colleague of Woese. He conceived a novel strategy to solve this dilemma. Pace adapted a technique for gene copying, or cloning, so that it could be applied directly to a community of different microbes in a sample taken from the environment. The technique, PCR, allowed him to make many copies of the DNA from all the microbes in the sample, obviating the need for growing the organisms. Pace believed he could then assess the variety of different types of microbes in the sample by analyzing the ribosomal RNA genes using Woese's techniques.

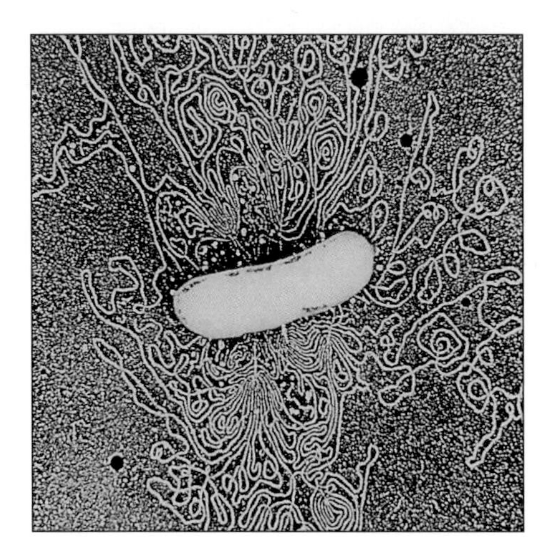

Breaking open a bacterial cell frees its DNA.

Pace collected samples from the dense, pink microbial mats growing in a hot spring in Yellowstone Park to try out his new approach. He was astounded by what he found. Instead of the few varieties of ribosomal RNA he expected, he discovered *hundreds*. This meant there must be hundreds of hitherto undiscovered species living together in a complex microbial community.

As Pace and other scientists studied more and more of the world's natural habitats, they duplicated his findings from Yellowstone. In each new habitat, there were new microbes. Scientists found them in the coldest and hottest environments and in the most highly acid or alkaline settings. They found them living in rock deep within the earth's mantle. They found them miles beneath the ocean's surface, under the earth's crust. There was virtually no place so extreme that complex communities couldn't be found—microbes occupied virtually all of earth's nooks and crannies. For the scientists exploring earth's habitats with this new approach, this was a bit like opening the door and stepping out onto a new planet. It also confirmed their suspicions that they had only been studying the tip of the microbial iceberg. We now believe there may be as many as 1,000 different kinds of microbes in the environment for every one that has been coaxed into growing in a laboratory. That translates into as many as 5 to 8 million different kinds waiting to be discovered.

Pace realized that he could produce beakers full of DNA and analyze it directly without having to grow the microbes in the laboratory.

It wasn't just the number of different kinds of microbes that was surprising, it was also their estimated sheer weight. Scientists now know that the microbial world is by far the earth's largest biomass. Microbes account for 95% of the biomass in the ocean, and their total biomass exceeds that of all the plants and animals combined.

Assigning microbes that couldn't be grown in a laboratory to a spot on the tree of life might not seem important when it comes to understanding the world in which we live. However, once scientists assigned a microbe to a specific group on the tree, they could tell a lot about what it might be doing in its natural setting. For example, knowing that a microbe was most closely related to a group that removes CO_2 from the atmosphere and releases methane gas, scientists could assume that the microbe's activities were to some extent similar.

The microbes that thrive under extreme conditions have become more than a curiosity. For one thing, we know that these microbes are tightly linked in the chemical web sustaining life in earth's biosphere. Gaining an understanding of their exact role is helping scientists better discern the way our biosphere works. Further, scientists are discovering microbes with properties that offer us alternative ways to solve some of our most difficult problems. We've found microbes that can clean up pollution in extreme environments by degrading the pollutants into benign chemical compounds. We've found microbes with unique enzymes that offer alternatives to the chemically driven processes of manufacturing, and microbes that produce an array of products with uses in everything from medical diagnostics to food substitutes to laundry detergents.

And thanks to the techniques developed by Pace and others, microbes don't have to be coaxed into growing in the laboratory for us to capitalize on their talents. Now scientists can simply copy their various genes directly from a spoonful of dirt or a cup of hot spring water.

A profusion of microbes

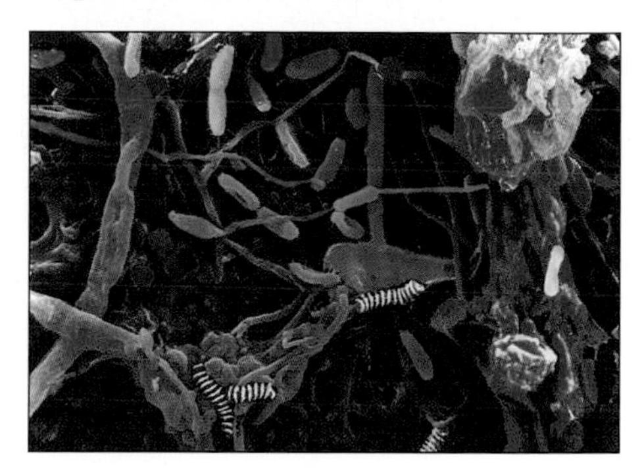

The number of different sequences of ribosomal RNA genes found in Pace's samples meant that there were an astonishing number of different kinds of microbes present.

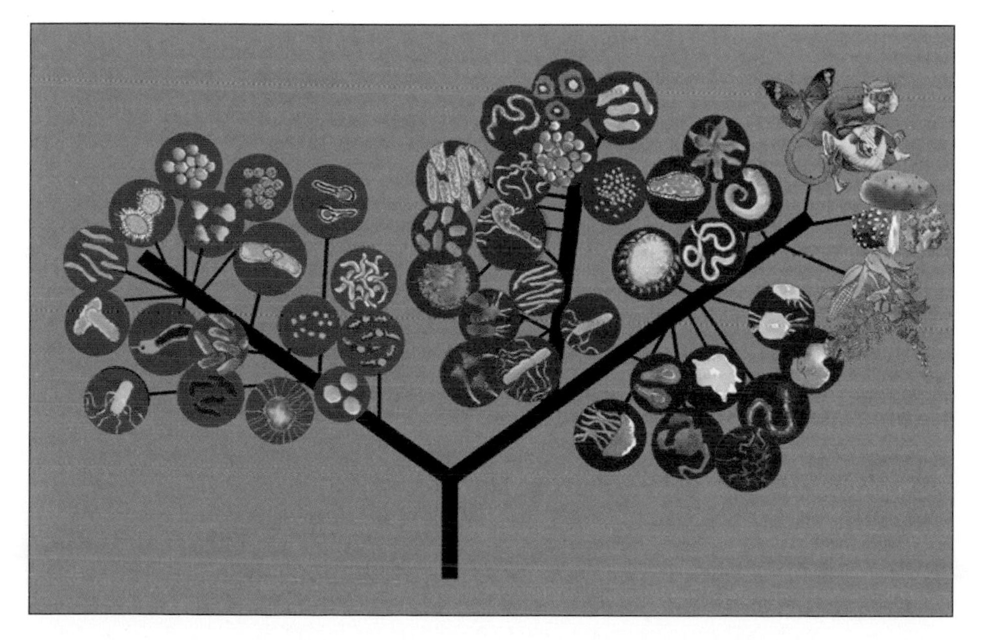

The discovery of all these new kinds of microbes necessitated adding many new branches to the ribosomal RNA tree of life.

Borrowing Life's Duplicating Machine

In most instances where scientists seek to use genes in experiments or to analyze their nucleotide sequence, they are confronted with too few molecules with which to work. It is essential as a first step to make many identical copies of the DNA in which one is interested. Until recently, the only way DNA could be copied was to make use of living cells. Bacteria are useful tools for this purpose: they reproduce rapidly and copy their own DNA each time they divide in two. And they can be induced to take up pieces of DNA from other creatures. They treat the foreign DNA as if it were their own, continue to reproduce, and consequently make as many copies of the foreign DNA as there are new cells.

The cells are then broken up and their DNA, including the foreign sequences, is isolated—usually millions of copies of it. This so-called recombinant DNA technology, discovered in 1973, is a relatively cumbersome procedure and takes several days to complete. Nonetheless, this important technology has contributed to many advances in our understanding of gene function in health and disease.

Scientists introduced a new technique in 1985 that allows for the automated copying of DNA without the aid of a living cell. The new technique relies on the bacterium's own DNA-making enzyme—DNA polymerase. The technique is called PCR, short for polymerase chain reaction, and it is now in widespread use. DNA from a large variety of sources, including prehistoric creatures preserved over millions of years, disinterred bodies, and tiny samples of blood, sweat, saliva, and semen, can be amplified to many millions of exact copies in just a few hours.

Gene copying now provides scientists with DNA to pinpoint genetic defects in persons at risk for genetic diseases and to detect the presence of bacteria and viruses from patients suffering from both common and obscure ailments. As we have already seen, gene copying also allows scientists to sample microbial habitats, revealing the unique genomes of hitherto unknown organisms.

Life's language uses four letters: A, T, C, and G.

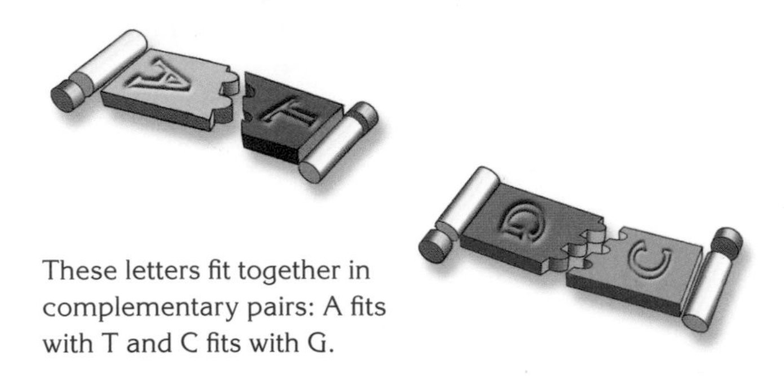

These letters fit together in complementary pairs: A fits with T and C fits with G.

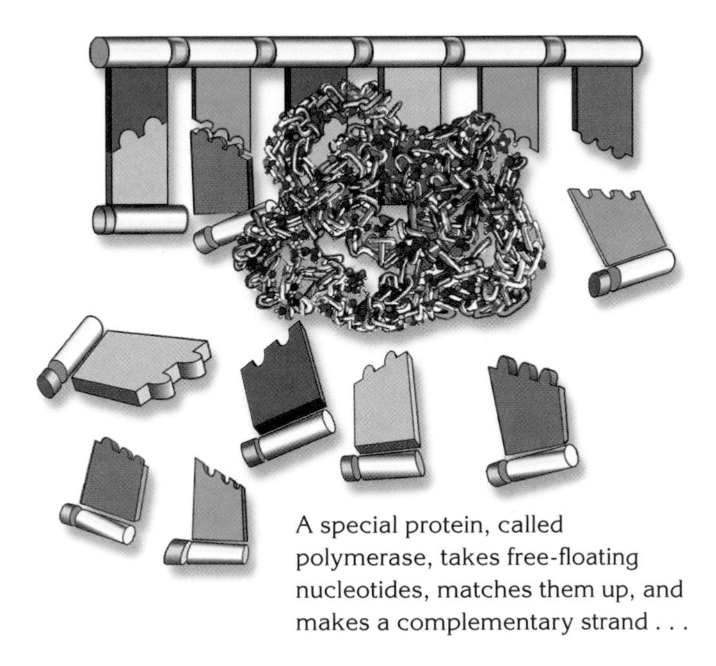

A special protein, called polymerase, takes free-floating nucleotides, matches them up, and makes a complementary strand . . .

. . . resulting in a molecule of DNA.

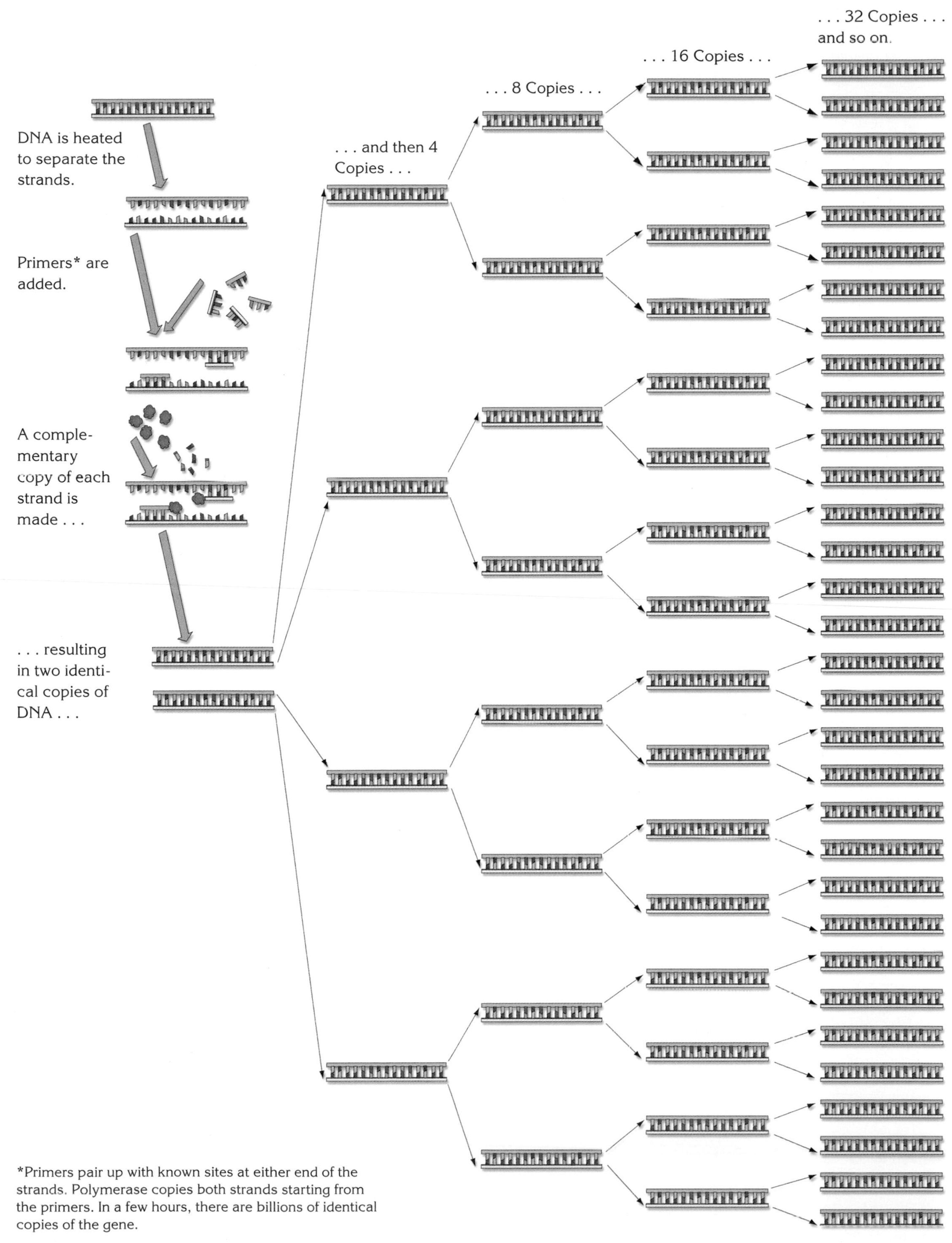

*Primers pair up with known sites at either end of the strands. Polymerase copies both strands starting from the primers. In a few hours, there are billions of identical copies of the gene.

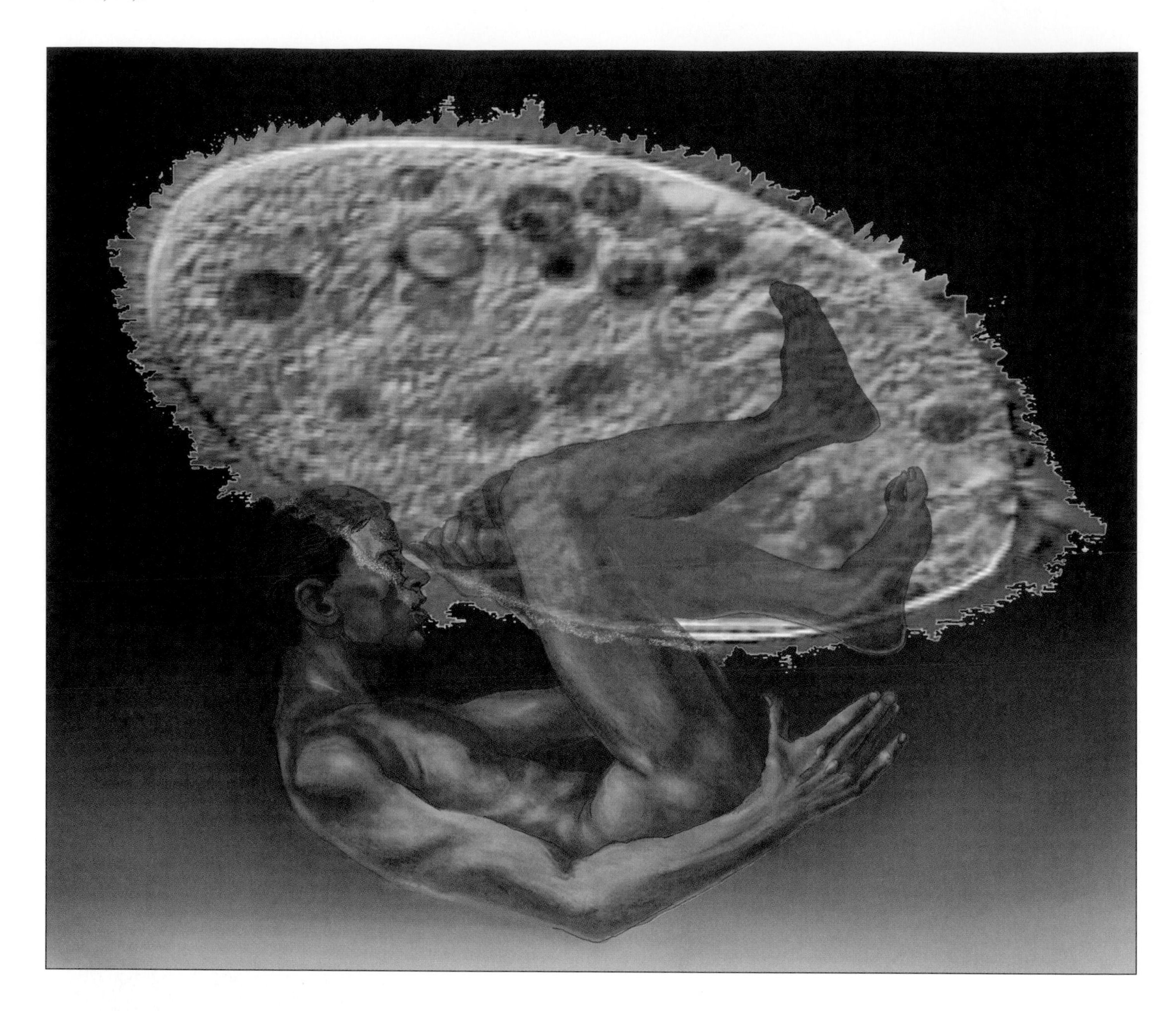

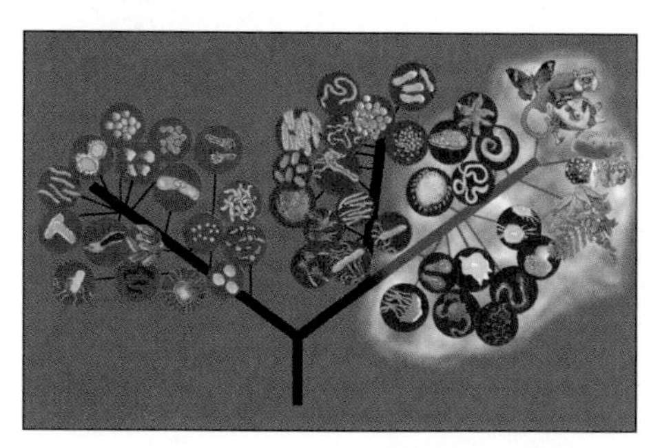

The Eukarya—larger cells with a true nucleus (which include animals, plants, fungi, yeast, and protozoa)—branched off from the Bacterial and Archaeal tree some 2 billion years after life got started.

Our Branch on the Tree of Life

Woese's breakthrough helped bridge an enormous gap in evolutionary history. Now, by following the relationships mapped out on the genetic tree, many scientists believe that the Bacteria and Archaea separated from their common ancestor very early in evolutionary history, forming two of the three domains of life that exist today.

Sometime very early on, a different kind of cell appeared. This cell was considerably larger than its bacterial and archaeal predecessors, and it had an important distinguishing feature. Unlike the Bacteria and Archaea, this larger cell had a nucleus, an internal membrane sac in which its DNA was stored. The new cell, called a eukaryote for "true kernel," is the ancient ancestor that gave rise to a third major domain—the branch of the tree we share with all other visible forms of life.

How and why the membrane-enclosed nucleus arose remains a mystery. We can, however, speculate that such an arrangement may have provided certain advantages over the primitive eukaryotic cell's non-nucleated counterparts. As life evolved and cells became more complex, the amount of DNA required to provide instructions increased. The increase in DNA must have made increasing demands on the cell's machinery to keep it organized and free from disruption by forces both inside and outside the cell. In modern cells, membranes serve as organizers, keeping various functions separated and running smoothly, and an internal membrane may have offered a partial solution to the primitive cell's DNA management problems.

Further, as cells increased in complexity, it meant there were more different kinds of proteins needed to do the cells' work. Since membranes serve as organizers, the increase in complexity was likely to enhance the value of more membrane space to keep the proteins organized and functioning smoothly. Soon, the more complex cells were larger and had more outer membrane, which proceeded to wrinkle and insert itself into the cells' interior spaces. At some stage, the cell's DNA molecules and their associated proteins may have become encircled by the evolving membrane systems, just as many other internal compartments were forming.

Whatever its origins, the nucleus remains a constant feature of almost all eukaryotic cells today, including our own. Once the new design emerged, it provided a welcome mechanism for compartmentalized genetic material that has carried forward to the present.

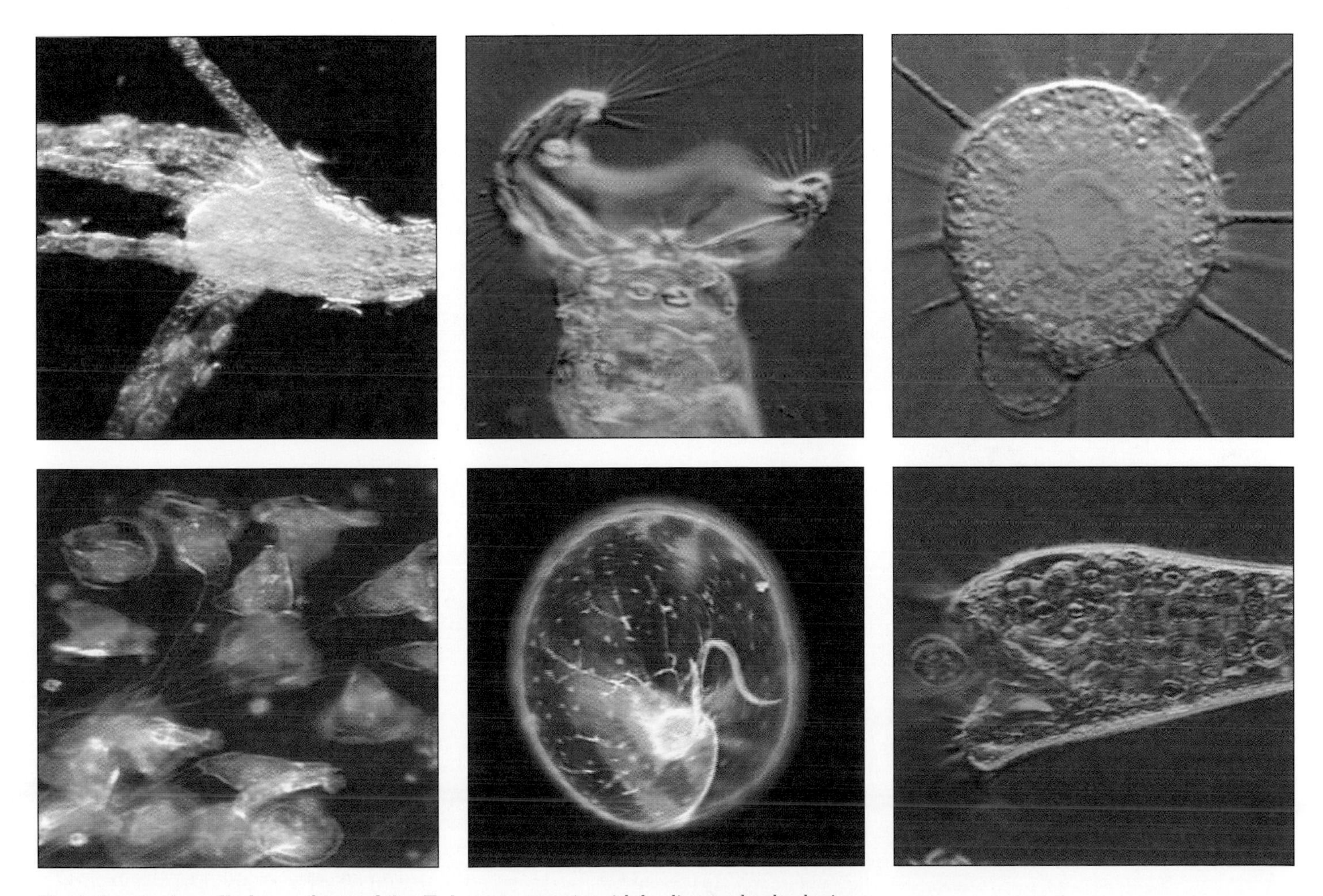

Even the single-celled members of the Eukarya come in widely diverse body designs.

Marriages of Convenience

Despite the advances in our basic understanding of how life progressed on the planet, we are still left with many questions about how the seemingly complex world of visible eukaryotic life evolved from the early world of Bacteria and Archaea. After all, the unseen world of microbes seems very far removed from plants and animals.

Somewhere along the way, our single-celled eukaryotic ancestors invented a number of abilities that distinguish them from their prokaryotic relatives. They invented true sexual reproduction, which provided a mechanism for mixing genetic information between two like cells. They invented a means of linking up with each other to form multicelled organisms, and they evolved ever more sophisticated systems for communicating with each other. This made it possible to develop a profusion of new body styles, rapidly outstripping the limited number of shapes shared by the Bacteria and Archaea.

Going Nuclear: A New Kind of Cell

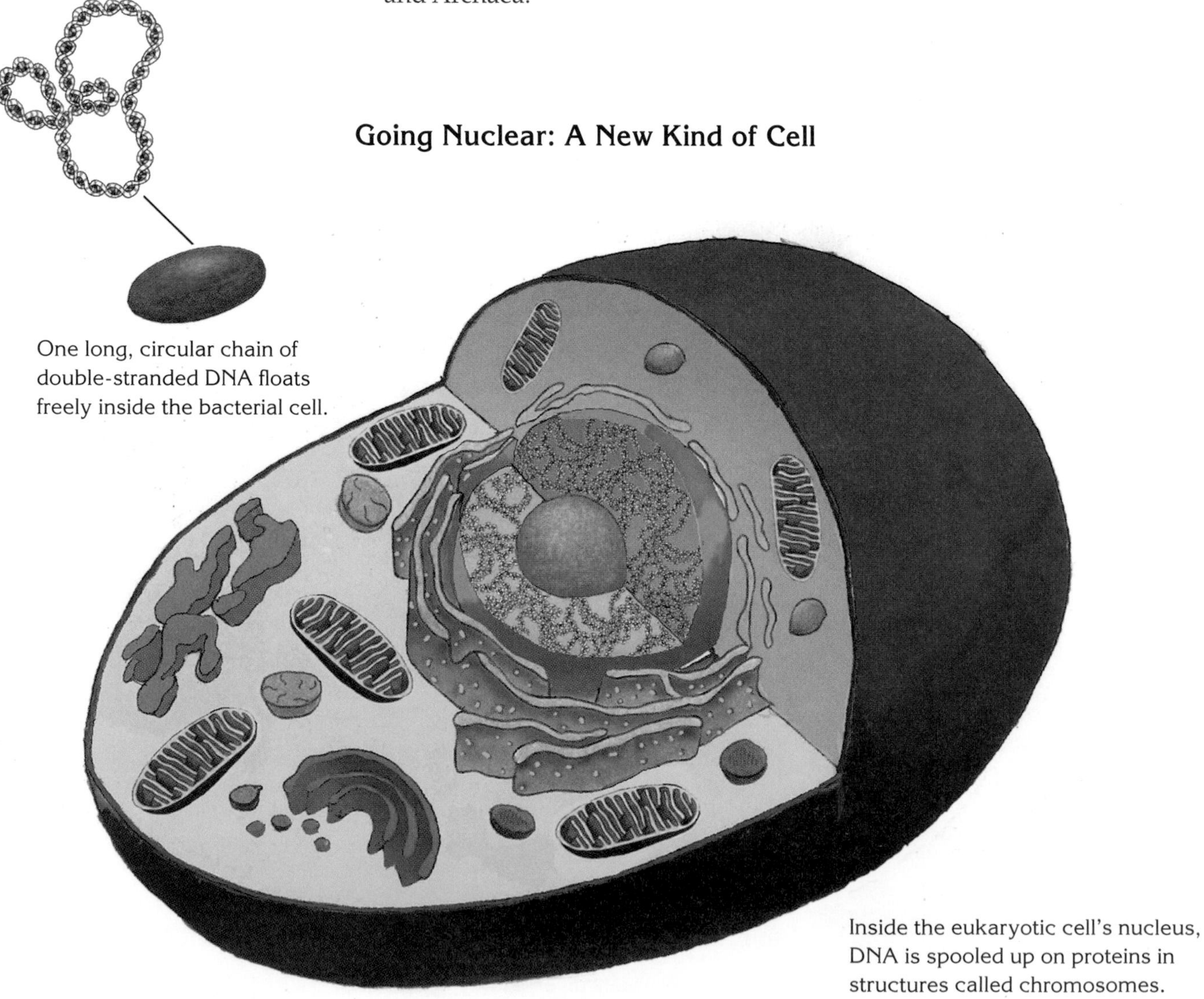

One long, circular chain of double-stranded DNA floats freely inside the bacterial cell.

Inside the eukaryotic cell's nucleus, DNA is spooled up on proteins in structures called chromosomes.

The great explosion of visible, multicellular life forms on our branch of the tree didn't really get started until 500 to 600 million years ago, a short span of evolutionary time. So, relatively speaking, we are very close evolutionarily to all of visible life today. Our kinship with chimpanzees is well known, but we also share at least 90% of our ribosomal RNA gene with all other animals, plants, and fungi.

In addition to the acquisition of a nucleus, there have been other startling leaps in internal architectural complexity that have puzzled scientists. The slow, step-by-step mutation and natural selection that had dictated the events of the preceding 2 billion years couldn't easily account for the major internal changes. The changes were so dramatic that they would be the biologic equivalent of taking "Chopsticks" and modifying it into a Beethoven symphony overnight.

Scientists began to find clues to this riddle in the late 19th century as newer and more powerful microscopes revealed more and more about the internal structure of plant and animal cells. Among the first things that became obvious was the striking resemblance between bacteria and certain structures inside eukaryotic cells.

One of these bacteria-like structures is found in almost all modern eukaryotic cells: a specialized organelle called the mitochondrion. Mitochondria function as the cell's essential energy factory. Our digestive system breaks down all the food we eat into smaller and smaller units that wind up inside our cells' mitochondria. The mitochondria then make usable chemical energy for our cells from these small food units.

Plants, algae, and certain protozoa all have a second bacteria-like organelle called a chloroplast. A chloroplast looks much like the mitochondrion, but it has a very different function. It contains chlorophyll, captures energy from sunlight, and converts it into food in a process called photosynthesis.

These microscopic observations prompted a German biologist in the late 19th century to advance a then controversial theory—that the organelles had once been free-living bacteria that had taken up residence long ago inside a larger cell. Biologists in the early 20th century further hypothesized that, once amalgamated, the two cells became dependent on each other, with one supplying chemical energy to its host and the other providing shelter and a steady diet to its resident.

The theory seemed pretty far-fetched at the time, and was slow in gaining support. However, as our knowledge about the way cells work continued to advance, evidence supporting the bacterial origin of these subcellular structures mounted. Today, this theory is well established.

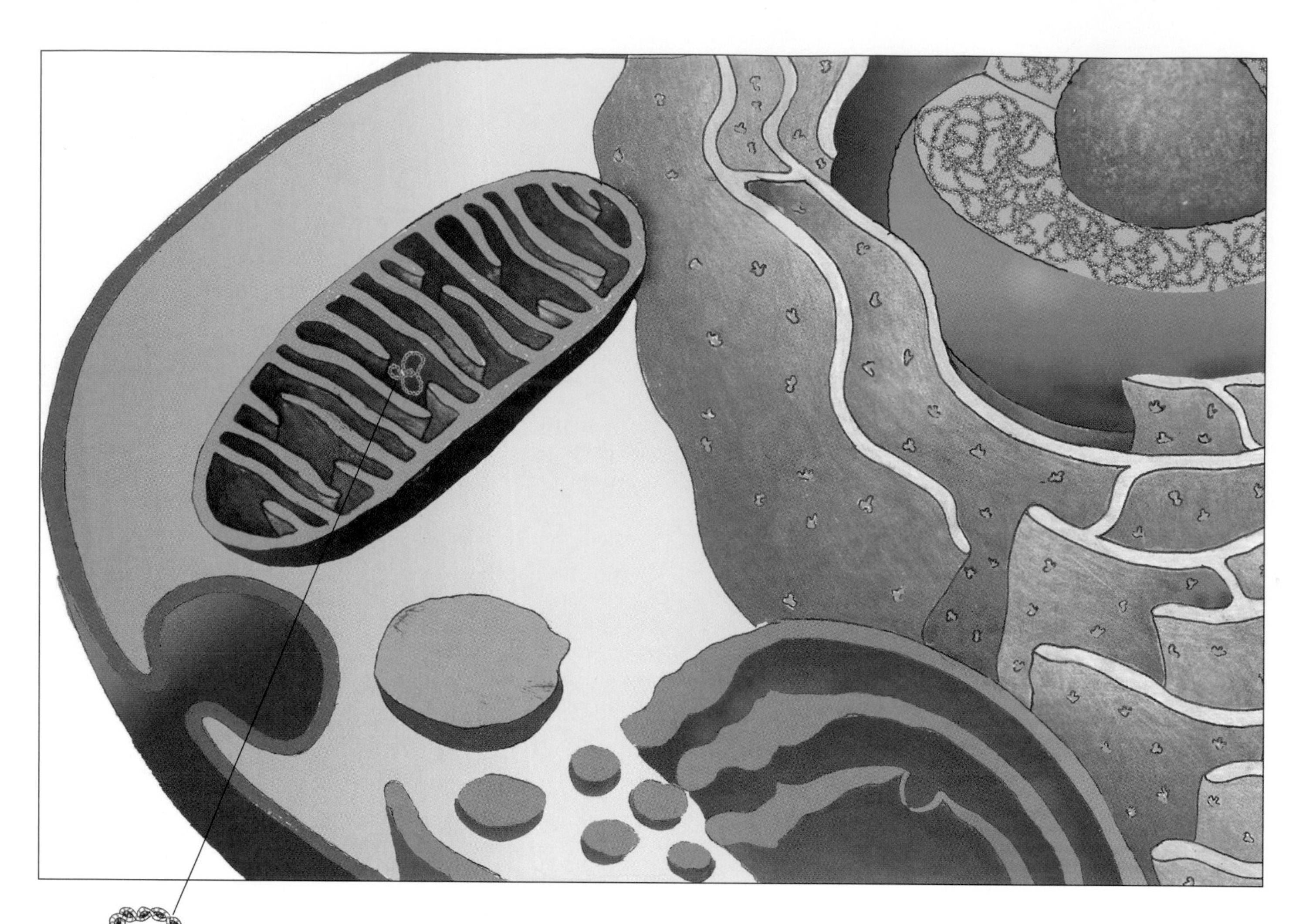

Mitochondria have DNA as well. It is a circular double-stranded chain similar to that found in bacteria.

Part of the scientific evidence for the bacterial origins of organelles was based on a closer look at the organelles' physical structure, and part was based on their biologic properties. Mitochondria and chloroplasts simply look and behave a lot like entrapped bacteria. They have their own DNA, which is very different from that found in the cell's nucleus. The DNA in a mitochondrion or chloroplast is a single, double-stranded circle, just like that found in bacteria. Mitochondria and chloroplasts have their own cell machinery—their own RNA and their own ribosomes for making proteins. And they reproduce independently from the cell by pinching in the middle and dividing into two new organelles, just as bacteria do. This activity is, of course, precisely coordinated with the cell's own division, so that the number of mitochondria per cell remains constant.

Modern chloroplasts even bear an uncanny physical resemblance to the photosynthesizing cyanobacteria, a highly successful group of microbes that occupy virtually every spot on the planet where there is sunlight.

How the Mitochondrion Got Its Home

How could such an extraordinary amalgam of two independent life forms come about? In the case of the mitochondria, scientists speculate that highly predatory bacteria may have invaded certain types of cells and fed on their insides—like space aliens. There are modern-day bacterial practitioners of this lifestyle. A microbe called *Bdellovibrio* attaches to its bacterial prey and turns itself into a high-speed drill, whirling furiously and punching through to the inside. It then essentially devours its prey from the inside out, leaving behind a dead and empty shell.

A more likely predecessor of at least some mitochondria is an ancestor of the bacteria called *Rickettsia*. Swedish scientists comparing the DNA in the eukaryotic mitochondrion of yeast with DNA in the modern version of this bacterium found too many similarities in the gene sequences for pure coincidence. The *Rickettsia* are even totally dependent for their reproduction upon infecting a cell. They have given up, among other things, the ability to manufacture the basic building blocks of their DNA. *Rickettsia* rely on their host cell for these essential substances, and they may represent an intermediate step on the way to becoming a permanent resident.

The suspected aggressive lifestyle behind the mitochondrial acquisition and merger might well have produced short-term success for the predator bacterium, but before long it would have run out of cells to invade. Some accommodation must have taken place. Perhaps the processes of mutation and natural selection began to modify the relationship. Maybe the prey adapted, becoming less susceptible to the lethal effects of the predator; maybe the predator adapted, becoming less virulent. Perhaps both. At any rate, what resulted is a true marriage of convenience—one in which the invaded cell gets the energy produced by the invader, and the invader gets a steady food supply and safe haven. Almost every eukaryote today, including humans, contains mitochondria and is thus a descendant of this long-ago marriage.

Mitochondrial Origins

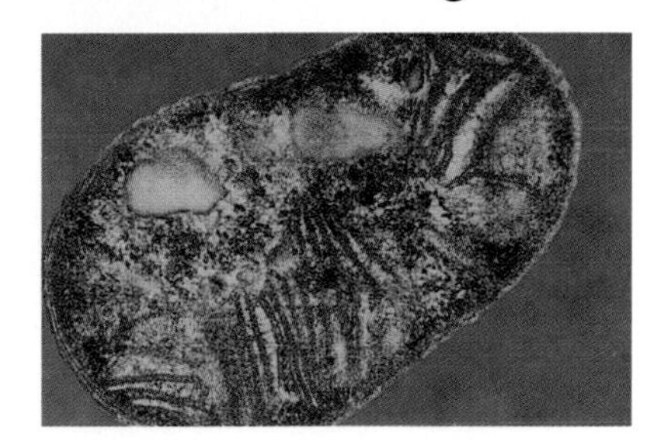

Identity: Formerly a bacterium
Residence: Inside almost all eukaryal cells
Favorite pastime: Generating energy
Activities: A member of The Energy Generators, this microbe resembles the structures called mitochondria that are the energy factories for almost all nucleated cells today; its ancestors may have taken up residence in primitive nucleated cells early in the evolution of our branch of the tree of life.

From invading predator to organelle?

1. A parasitic bacterium invades a larger cell.

2. It is able to multiply inside, and begins to share the products of its metabolism with its host.

3. After many generations, invader and host have come to depend on each other, each dividing separately but in concert with each other.

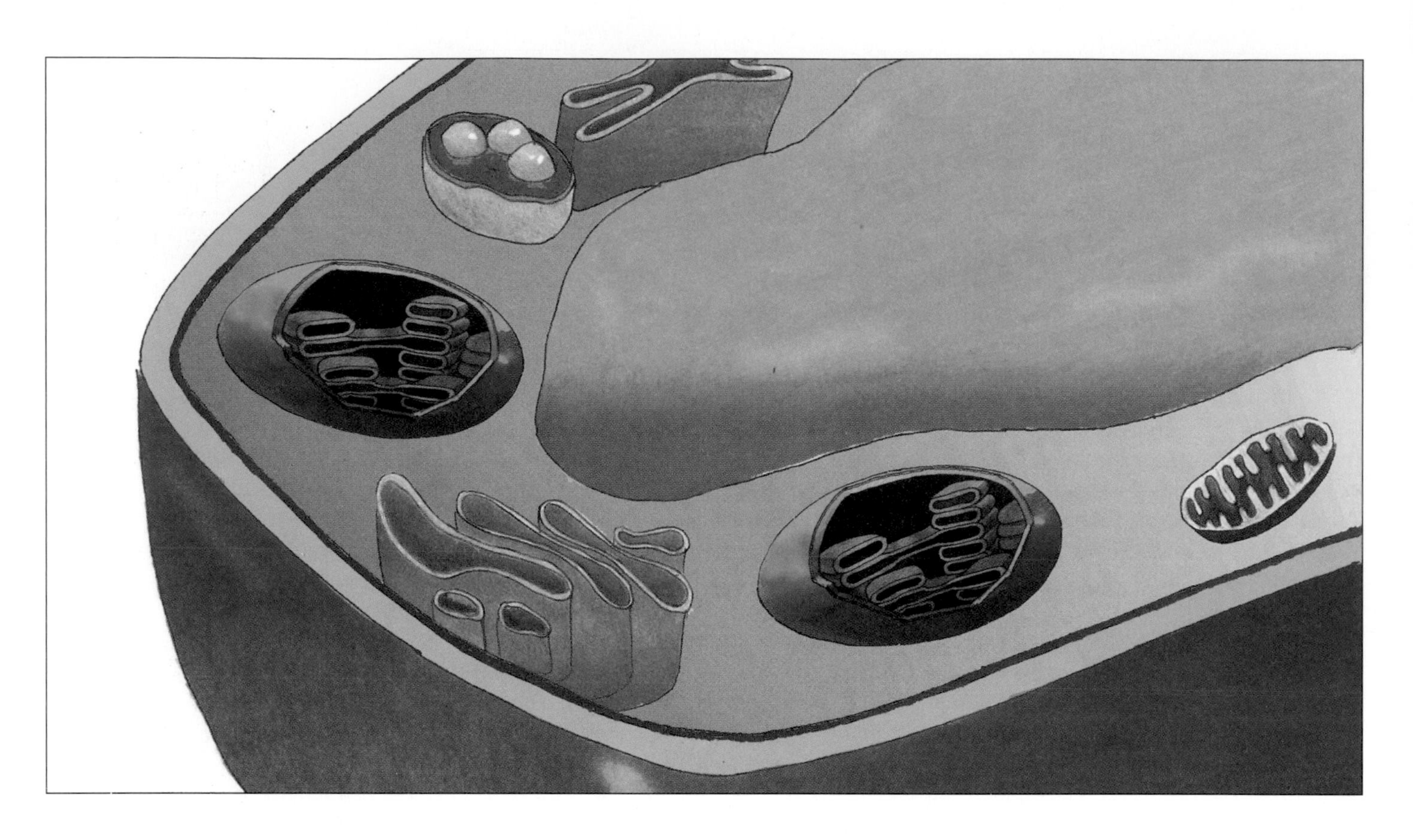

Synechococcus lividus

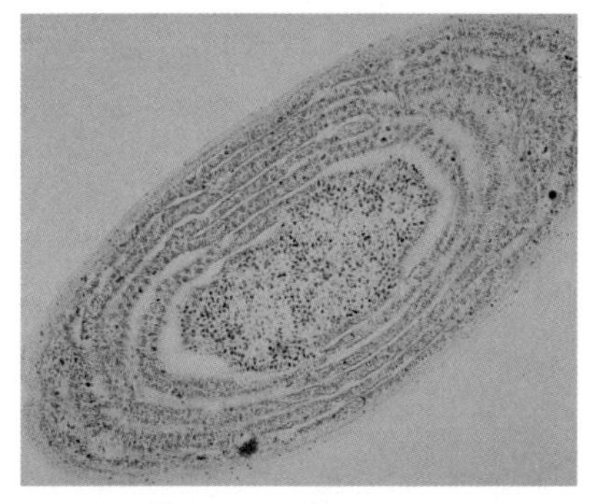

Identity: Bacterium
Residence: Water
Favorite pastime: Basking in the sunlight
Activities: A member of The Oxygen Generators, *Synechococcus* resembles the structures called chloroplasts that are found in plants today and are responsible for transforming energy from the sun into usable energy for the plant cell.

Mom, I Think I've Got Chloroplasts

The eukaryotic acquisition of the chloroplast ancestor happened sometime after mitochondria were incorporated. It involved a subset of the primitive eukaryotic cells that were destined to evolve a very different lifestyle. Thus plants, algae, fungi, insects, humans, and other animals all have mitochondria, but only the ancestors of the plants and algae, along with a small number of protozoa, got chloroplasts. This has made plants and algae self-sufficient, since they could now convert solar energy directly into food.

Scientists speculate that the history of chloroplasts is likely just the reverse of the mitochondria. They entered the host cell not as invaders but as food.

Perhaps photosynthesizing bacteria were the victims of an ancient feeding frenzy—ingested and, at first, only partially digested by larger cells. Imagine the delight of the first cell to gulp down the chloroplast ancestor, when its meal began to convert sunlight into usable food. At any rate, over millions of years of "indigestion," this also became a successful marriage, with every photosynthetic eukaryotic cell alive today a reminder of the union.

The incorporation of the chloroplasts and mitochondria into eukaryotic cells offers us a very different way of envisioning evolution. Cooperation seems to be as powerful as competition. Cooperative arrangements like the ones giving rise to the eukaryotic organelles played a decisive role in the evolution of multicellular life forms.

A Modern Tale of Microbial Marriage

Kwang Jeon was feeding his amoebae when he noticed that something was amiss. Amoebae are a little bit like miniaturized versions of The Blob, and Jeon had been studying them in his laboratory at the University of Tennessee for the past several years. All of his amoebae had lost their characteristic slithering motility and had changed their shape. They looked quite ill.

He had recently received a new shipment of the little one-celled protozoans and had divided them into containers lined up with those containing his other amoebae collected from around the world. As the days went by, almost all of his amoebae died. Jeon was horrified when he realized the cause. The newly arrived amoebae had apparently brought a bacterial infection along with them, and they had subsequently infected all his other containers. When he looked at these suffering protozoans, they all contained huge numbers of the pathogenic bacteria. He thought that all was lost.

However, he was wrong. A small number of Jeon's amoebae survived. The survivors of the epidemic weren't particularly healthy. They only divided once a month, rather than once every two or three days as his healthy amoebae did. They were very sensitive to little changes in temperature and food, and they were still infected with bacteria, just not so many.

Jeon decided to try to cure his struggling amoebae of their infection. When he put an antibiotic that only killed bacteria into their food, however, the amoebae died along with the bacteria. Over the next 5 years, Jeon continued to nurse these creatures. He selected the amoebae that were stronger and let the others die off, and he eventually wound up with "bacterized" amoebae that were normal except for their microbial inhabitants.

Through a series of experiments, Jeon was able to show that the bacteria couldn't live outside the amoebae and the amoebae couldn't live without the bacteria. It's possible that Jeon witnessed something similar to the evolutionary event that happened 1.6 billion years ago when the first experimental models of eukaryotic cells were being designed.

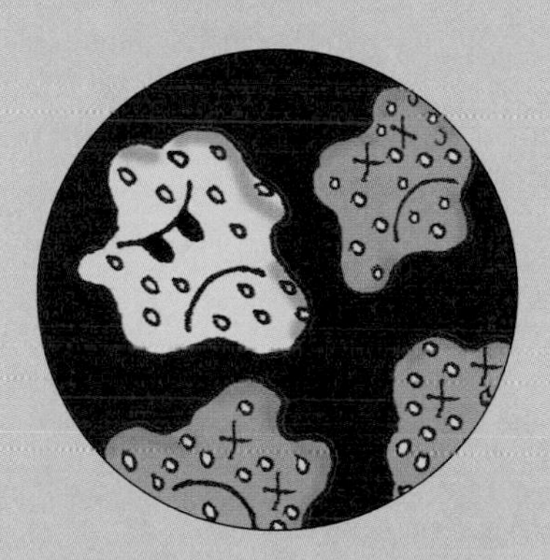

Jeon's original amoebae . . .

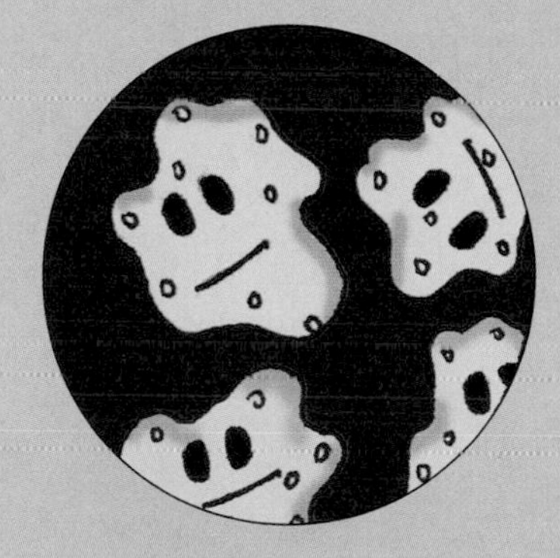

adapted over generations to their bacterial invaders . . .

until both came to depend on each other.

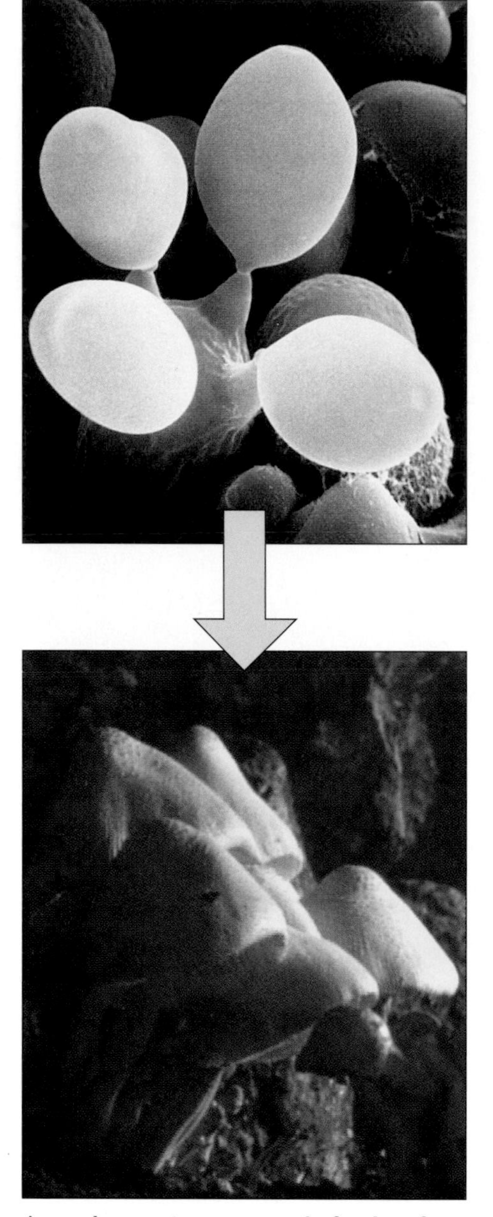

A mushroom is composed of only a few kinds of cells, and each has a different function.

From One to Many, From Many to One

Among life's great, unanswered questions is how multicellular life forms evolved from single cells. Unlike creatures that consist of a totally self-contained single cell, larger life forms are composed of many different types of cells, each with a different function. Although it is unlikely that we will ever resolve the sequence of events that led to multicellularity, many scientists believe there are clues lying within the microbial world.

Consider the slime mold. When food is plentiful, the mold exists as individual cells in the soil. When food becomes scarce, the cells signal each other and come together to form a mass that is readily visible to the human eye. More important, cells within the collective begin to assume different functions. Some cells differentiate and form a stalk. Other cells form spores, a form of the organism that can survive until conditions improve. The chemical signals passing back and forth among the individual cells tell each what their final role will be. When we look at the development of multicellular creatures, a similar chemical process dictates whether a cell will be a skin cell, nerve cell, and so on.

We usually think of mushrooms as the delicate structures that appear in the forest and on our lawns after a rain or in the vegetable section of the supermarket. But mushrooms are fungi, existing for most of their lives as tiny cells arranged in threadlike structures called hyphae. Each individual cell of the threadlike structure is a self-contained unit, and hence mushrooms are included with the microbes. Only at certain times during their lives do the hyphae migrate together and differentiate into a collection of cells with different functions much like the slime mold. The purpose is reproduction, and the end result is the beautiful and visible structure that we call a mushroom.

An even more intriguing set of clues exists within certain types of microbial communities. Composed of a variety of different cooperating species, these miniature collectives of individual cells function in some ways like a multicellular organism. Different species within the community perform different tasks. Some take food directly from the environment and convert it into compounds that other members of the community can use; others form a gluelike substance that holds everyone together in one place, and so forth.

The members of these collectives often become highly interdependent. Thus if one member is removed, the community functions less well as a whole. If several members are removed, the community may not survive. Perhaps it is not so hard to imagine that early collectives of cooperating cells like this became inseparable, laying the groundwork for the vast array of multicellular creatures that exist today.

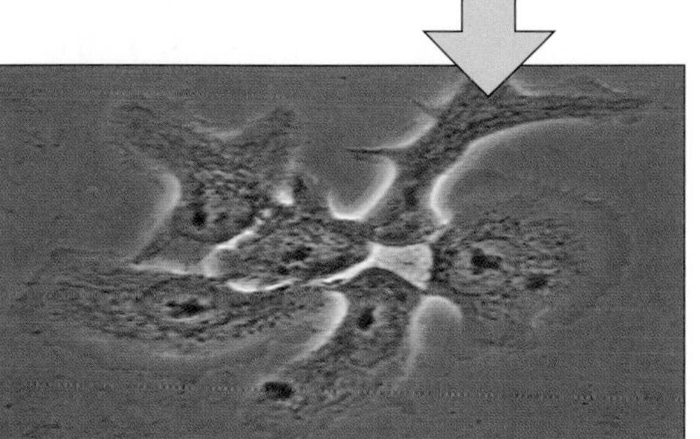

A penguin is composed of hundreds of different kinds of cells, each kind performing a different function.

Evolution on Fast Forward

Swimming in DNA
Microbes have the unique ability to capture free-floating DNA from their environment. Such a practice can lead to the acquisition of new genetic traits.

What other revelations lie within the DNA? What more can we learn about life's origins and evolutionary progress by reading the genetic histories of living creatures today?

DNA technology has advanced dramatically from the time Carl Woese began his search. Today, scientists have turned their attention to far more than just the single ribosomal RNA gene.

Some of the information being generated has confirmed the view of the tree of life based on the ribosomal RNA gene. But other information has offered a surprise.

Trees of Genes

As scientists compared increasing numbers of genes from increasing numbers of microbes, a puzzling picture emerged. Scientists found that constructing trees using certain other genes yielded very different trees of relatedness, none of which was in agreement with the others. This was enough to give everyone pause. How could this be?

We have generally assumed that genes are transferred vertically—from one cell to its daughter cells, from parent to offspring. If this assumption were true, all of the genes in the genome of a species would have evolved in parallel. Scientists should be able to build family trees from *any* shared gene in the species' genomes with identical results.

So why did comparing some genes produce a totally different tree? The answer may rest in a phenomenon that occurs commonly in microbes called horizontal gene transfer. Microbes can pass genes directly to their neighbors! This is the microbial equivalent of a human acquiring a genetic characteristic from a friend or coworker through a handshake. The process is clearly distinct from vertical gene transfer, which is the passage of genes to progeny during cell division.

A microbe accomplishes horizontal gene transfer by taking in DNA from another microbe, thereby gaining new genes and consequently new capabilities. There are a variety of strategies for this, and microbes have apparently been doing it well back into evolutionary history.

Microbes' use of both horizontal and vertical gene transfers may explain the conflicting results in constructing genetic trees of evolutionary relatedness. Individual genes in a microbe's genome might have different lineages. A tree built from an individual gene would consequently show the path of that gene, but not necessarily the path of the microbe's evolution. If all of a microbe's genes could be transferred horizontally, then each gene tree might thus have a very different array of branches from any of the others.

Vertical Gene Transfer

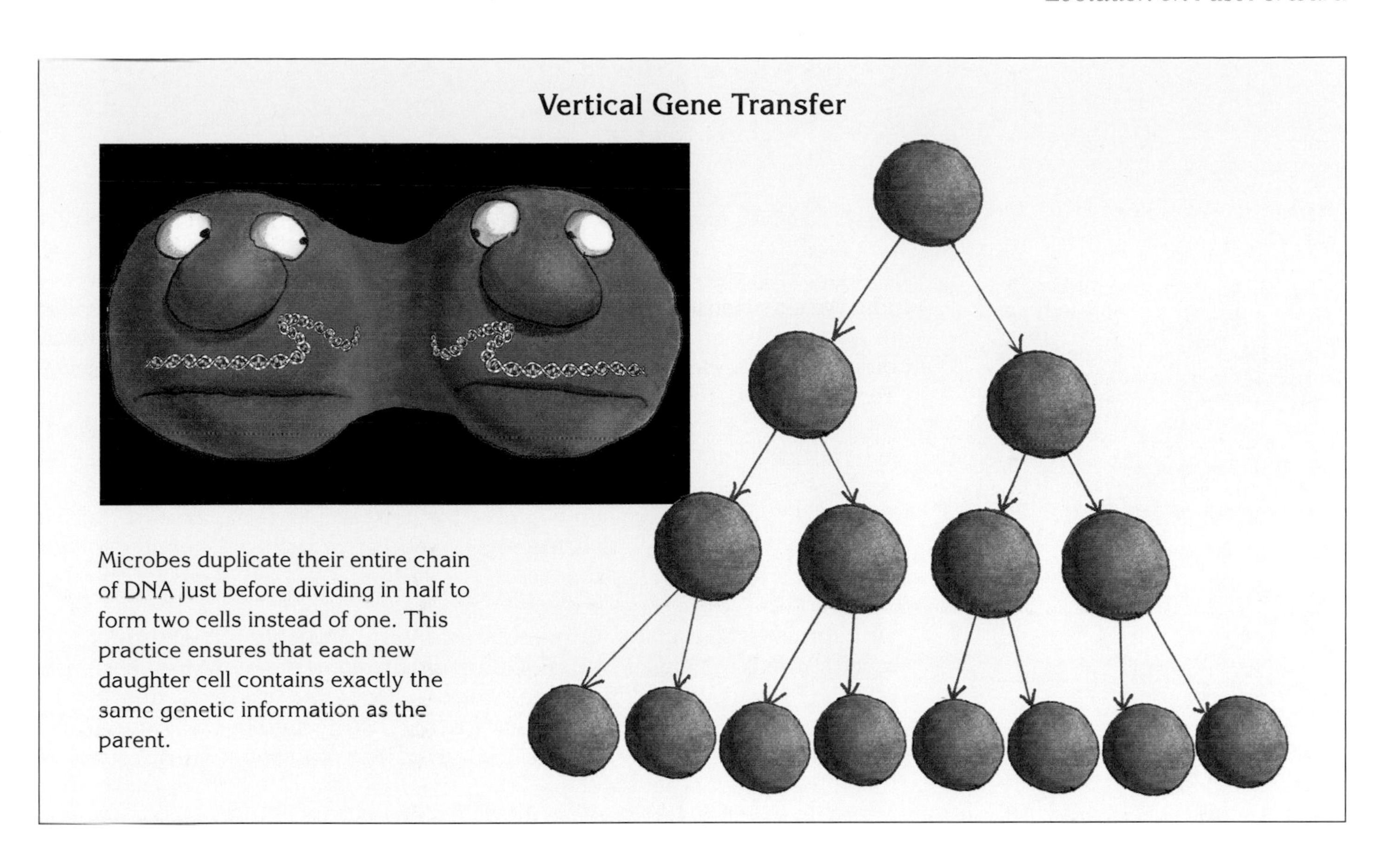

Microbes duplicate their entire chain of DNA just before dividing in half to form two cells instead of one. This practice ensures that each new daughter cell contains exactly the same genetic information as the parent.

Horizontal Gene Transfer

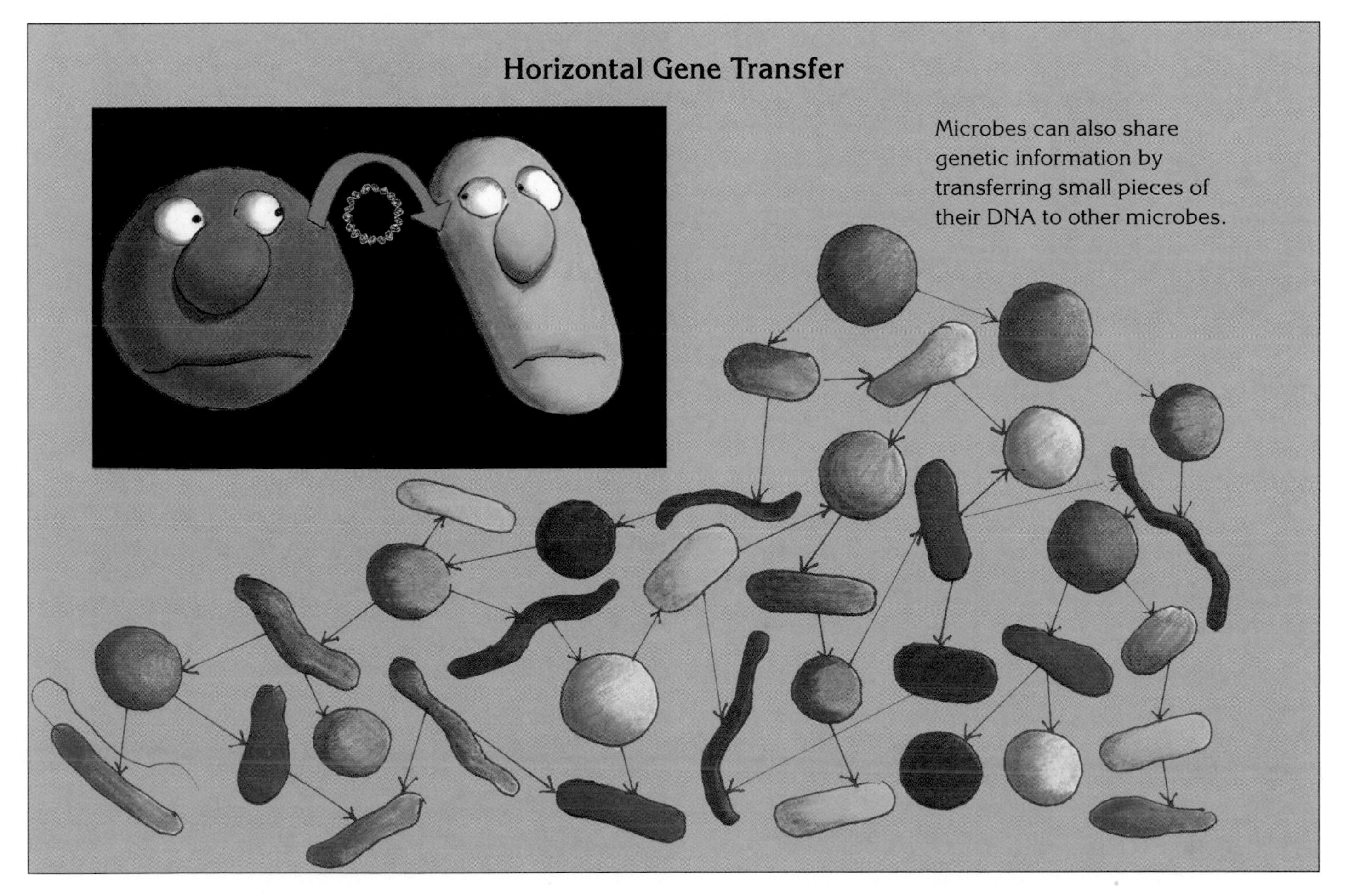

Microbes can also share genetic information by transferring small pieces of their DNA to other microbes.

The Great Gene Shuffle

This is not exactly the case. Some genes produce trees that are very similar to the tree based on the ribosomal RNA gene. Like the ribosomal RNA gene, these genes code for some of the biologic functions shared by all living cells that are likely to be at the very roots of life—the processes for making proteins and maintaining and passing along information to subsequent generations.

Such basic functions probably evolved in the very earliest cells—primitive life forms that existed long before the three main branches split and evolved along separate paths.

Horizontal gene transfer, quite independently, probably allowed microbes to take in entire gene cassettes—long stretches of genetic information containing instructions for multiple proteins. The DNA cassettes might contain, for example, the genes for proteins, that are needed to break down and utilize entirely new compounds, giving an edge to the recipient microbes when food sources get short.

Microbes can actively transfer certain types of genes very rapidly and to many individuals when the genes convey highly useful characteristics. Certain genes for antibiotic resistance have sped to microbes all around the globe in a matter of months rather than the hundreds of thousands of years that might be predicted if the microbes had to rely on mutation and selection alone.

Remarkably, the gene recipients don't even have to be from the same species. In fact, they apparently don't even have to be closely related. Consider the Bacteria and Archaea. We now believe that certain genes were traded back and forth between members of both domains. This would be about the equivalent in the visible world to a passing honeybee capturing the genes for flowering from a nearby tree.

If plants and animals swapped genes

With multicellular creatures, the practice of passing around genetic information just doesn't work.

It's no wonder that our view of evolutionary relationships became confused, and it's this revelation that may once again change our view of how life progresses on our planet.

Gene sharing among the vastly diverse members of the microbial world gives them a tremendous advantage in rapidly adapting to changes in their environment. Members of the visible world, lacking the ability to transfer genes with such fluidity, must rely on the relatively slow processes of gene variation, heredity, and natural selection.

Gene sharing, and the ability to reproduce over a relatively short time, may be the two phenomena most responsible for the enormous biologic success of microbes on this planet. The microbial world was the first life on earth, and the extraordinary ability of microbes to adapt rapidly to changing circumstances could enable it to be the last as well. Whether we humans, with all our knowledge and intelligence, have the ability to accompany microbes into the future remains to be seen.

The Sequencing of *Thermotoga maritima*

T. maritima is one of the first microbes to have had its entire genome sequenced—its entire genetic score revealed.

Karen Nelson at The Institute for Genomic Research led the team that unveiled *T. maritima*'s genome. Once they finished, she and her colleagues thought that they were on their way to a major scientific breakthrough. Based on this microbe's genomic sequence, it initially appeared as though it might have existed before the Bacteria and Archaea split apart into their separate branches. This was an exciting finding, since it would have placed *T. maritima* closer to the ancestral cells present at the very beginning of life than any other microbe discovered so far.

But they were in for a disappointment. Although almost 25% of *T. maritima*'s genes are most similar to genes among members of the Archaea, its ribosomal RNA gene places it on the bacterial branch. Once they analyzed the genome in detail, a different picture emerged. *T. maritima* is very likely a prime example of evolution on fast forward. Many of the microbe's genes seem to have been acquired from a member of the Archaea. Even this was a pretty exciting revelation, since it provides scientists with evidence that massive amounts of gene swapping may have gone on between these very different groups of microbes.

Karen Nelson

Karen Nelson grew up on the island of Jamaica, surrounded by nature. "I'm driven by being able to answer questions that we could not answer five or ten years ago with the technology we had. We're sequencing the entire genomes in one go and to me that's just absolutely amazing."

Forward to the Past

It's perhaps difficult for some of us to think of microbes—lowly "germs"—as our distant ancestors. They are, after all, so different from us on the surface. Yet, as we trace our family tree all the way back to the beginning of life on earth, we become increasingly aware that all living things on this planet share many common features. The footprints in our DNA lead us back invariably to the very first successful life form on our planet—to single-celled creatures that continue to dominate the earth today.

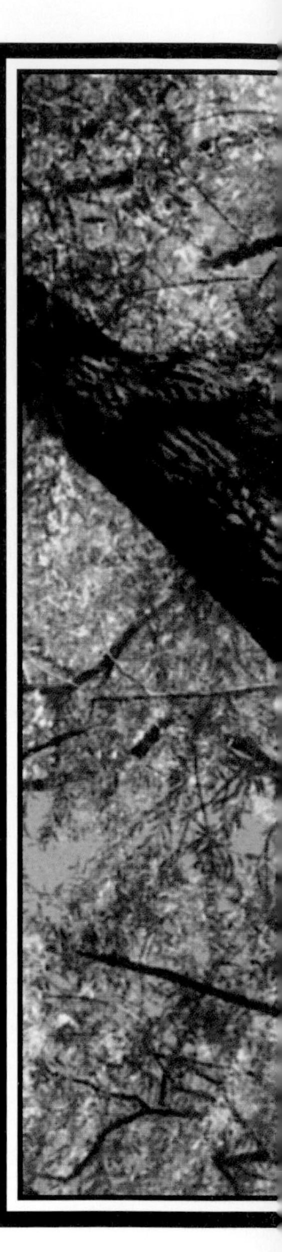

It may be even more difficult to view microbes as the most successful life forms, but despite their simplicity of design, they are unique in their ability to exploit every conceivable niche that earth has to offer. Microbes invented much of what exists today. For 3 billion years microbes alone were life's experimenters, creating the basic processes of metabolism, motion, reproduction, sex, and communication. And they will continue this role far into the future.

We humans are but a brief riff in the immense symphony of evolution—a symphony performed exclusively by microbes until very recently. We are, however, beginning to play a unique and more conscious role in the direction of the evolutionary process. Biotechnology, after all, is little more than our attempt to emulate the finely honed skills of genetic manipulation that microbes have been using for at least 3.5 billion years.

What will be our future success? Our efforts may be more directed, but theirs offer a greater range of experimentation. While we worry about a new transgenic corn strain, they have created, tested, and discarded millions of possible new life forms.

We trust that we will show proper concern that we might take the process in unintended and unhealthy directions. But what we fail to recognize is that from the evolutionary perspective, we are but a finger cymbal in the orchestra, and if our efforts to improve the world go awry, we may find ourselves among the extinct species.

Evolution will continue with or without us. And life will continue to do what it does best—adapt and evolve.

SECTION THREE

DANGEROUS FRIENDS AND FRIENDLY ENEMIES

Disease usually represents the inconclusive negotiations for symbiosis . . . a biological misinterpretation of borders.

—Lewis Thomas

Suzanne Kelly woke on Friday morning and knew something was terribly wrong. "I felt incredibly heavy, like I was made of cement. It took so much energy to move and breathe. By evening, my fever was 102.5 and I ached so badly that I thought I was getting the flu." Suzanne was a victim of a mysterious new disease that emerged in the Four Corners area of the southwestern United States. Suzanne was one of the fortunate. She survived.

On May 9, 1993, the disease took the life of a strong, healthy, young Navajo woman—an athlete. Five days later, it took the life of her fiancé, also an athlete. And then it took four more lives.

Local physicians realized they were seeing something new. The disease was quick, leaving its victims gasping for breath, sending them into cardiac arrest within the space of hours. And it killed half of its victims. The number of cases continued to climb—a mother of two, a man repairing his house, a teenager. Panic spread, first among the people of the Navajo nation and then in the surrounding communities.

The alarms went off, bringing in the nation's disease hunters from the Centers for Disease Control and Prevention (CDC) in Atlanta, Georgia. C. J. Peters, Director of the Emerging Pathogens Branch, remembers.

"We got the call from the folks in the Southwest. They had reached a point when they knew they needed help. We came out in force. We didn't know what this was. We knew people were dying, we knew we had a problem with some kind of disease. It looked like an infectious disease, it had a very high mortality rate."

CDC investigators moved into the area to set up their investigation. They sent blood and tissue samples to the CDC laboratories in Atlanta, where they were screened for toxins and all the known causes of disease.

C.J. Peters

On June 3, the CDC investigators got lucky. The laboratory found evidence of hantavirus infection.

This hantavirus was far more lethal than its cousins in Europe and Asia. The Four Corners hantavirus filled the lungs of its victims with fluid. The fluid led rapidly to lung collapse and, in many cases, death.

The epidemiologists found the source of the virus—the deer mouse, a familiar denizen of the desert. For the mouse, the virus was innocuous. For humans coming in contact with the infected mice, the virus was lethal.

Members of the public health community know the devastation that a widespread epidemic can wreak. The influenza pandemic in 1918–1919 killed at least 21 million people worldwide. The AIDS pandemic continues to take its toll, with millions dead and millions more infected. Fortunately, this hantavirus outbreak was small and remained local. Through the combined efforts of the regional and national public health services, the outbreak was stopped, but only after 28 people were dead.

We live in a delicate balance with our microbial partners. Although we live in harmony with most, some relationships are unfriendly. Sometimes friendly microbes turn against us. Sometimes familiar enemies that we have learned to live with gain the upper hand. And sometimes swift, lethal strangers quickly overwhelm us. Although we can reach accommodation with some of our adversaries, we keep most at bay using public health measures and our most sophisticated defense—the immune system . . .

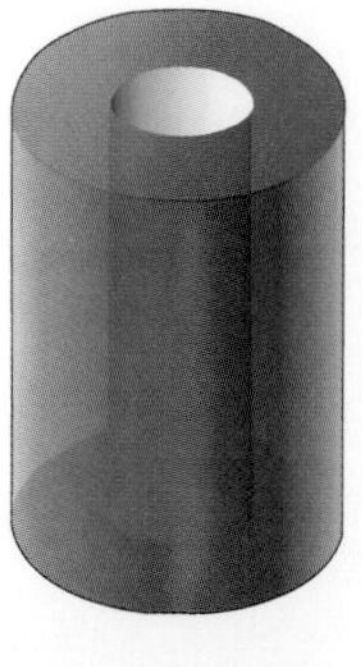

A tube has an "in side" and an "out side."

"In side"

"Out side"

Really inside

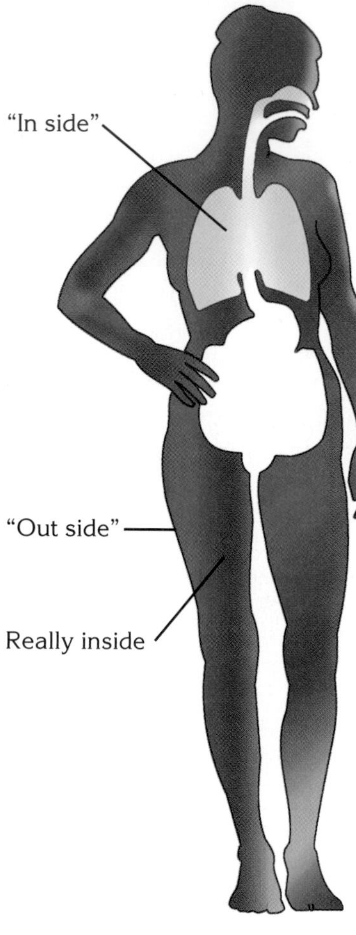

Microbes residing on our skin, mouth, and intestinal tract are on both the "out side" and the "in side" of our tube. Microbes invading muscle, bone, and internal organs are REALLY inside.

Living in a Microbial World

We are never alone. We have literally billions of microbes living in us and on us. But they aren't really in us. They inhabit our skin, our mouths, our digestive tracts—colonizing every surface that has access to the outside. In fact, from the microbe's perspective, we are a huge hollow tube, carrying them from place to place and keeping them nourished. These microbes are key players in our amazing, intricate, and dynamic personal ecosystems. We rely on these ecosystems—our "normal flora"—to keep us healthy.

Although our partnership with this normal flora is usually mutually beneficial, the relationship can turn adversarial. The natural drive to get more and do better is as common in the microbial world as it is in our own. From our viewpoint, we prefer that microbes stay in our hollow tube, outside the sterile tissues that make up our body. From their viewpoint, the grass may appear greener on the inside of the tube.

This difference of outlook can sometimes breed contention. Even the friendliest microbe turns dangerous if it breaches our skin barrier and relocates in our muscle, blood, or bone.

We have evolved a set of defenses to keep microbes on their side of the line. Our skin and mucous membranes form a barrier between what is really inside us and what is really outside us. Fluids bathing our body surfaces—like tears, saliva, and stomach acid—contain enzymes and other compounds harmful to many microbes. Cells, functioning like a well-trained border patrol, recognize foreigners when they step across the line from outside to inside. The border patrol cells link to a set of internal defenses that build better and better protection with each new encounter.

We each differ in our defensive repertoires and our ability to effectively employ them. Some people get influenza while their office mates don't. Our genetic makeup and many additional factors—nutrition, stress, possibly even our psychological well-being—can build or weaken our defenses.

Microbes have evolved their own strategies for living with us. The microbes in our normal flora have muted their invasive qualities and are generally held in check at our boundaries. Only rarely do they cross the line and turn into dangerous friends, and our defenses are exceedingly good at evicting them from forbidden spaces when they do.

If We Could Talk to the Microbes

"Microorganisms do communicate. They do it chemically."

A conversation with a microbe may seem like a strange idea. But that, say scientists like Stanley Falkow, a microbiologist at Stanford University, is exactly what happens. He and others speculate that over the centuries of microbe/human encounters, a set of chemical signals—silent conversations—have evolved. Our cells and the microbial cells both understand the signals, participating in a dialog complete with greetings, signs of recognition, and deceptions. Such conversations are likely the secret to the success of our longstanding enemies, like *Salmonella typhi*.

"Microbes have a way of talking to each other. They produce substances [that] each one can monitor. As the concentration of the substance increases, the pathogens know they have reached the proper population size, [and] they switch into a higher gear to become aggressive.

"The other thing that organisms do is [to] eavesdrop quite well. A lot of the pathogenic organisms have learned to understand how our cells talk to one another. One of the first things that many pathogenic organisms do when they make contact with our cells is cut the lines of communication to other cells in our body, so that they can't shout, 'We have an invader coming!' They simply silence them."

Other microbes, familiar enemies, use highly developed tactics—clever disguises that cover their surfaces and chemical messages that signal our cells—to outwit our defenses. They have traveled along with us for thousands of years and know us well. They routinely trick even the most sophisticated defenses that we can mount, allowing them to cross into our protected space and stay there.

At the extreme are the dangerous enemies—like the hantavirus—that we rarely encounter. Often they have evolved in association with other species. We are just accidental hosts to them—in the wrong place at the wrong time. Our defenses are not familiar with them, and the damage they wreak can be devastating.

We exchange thousands of silent messages with the inhabitants of this microbial world every day—chemical messages that say, "Glad to have you along," "Please come inside," or "Stay in your place." These cellular conversations distinguish friend from foe, familiar from alien. They precipitate action and counteraction.

Throughout our lives, we will participate in a dynamic relationship with this invisible world. Understanding that relationship can help each of us in both personal choices that may affect our own health and policy decisions that may affect the health of the world.

Stanley Falkow

Cell-Cell Communication

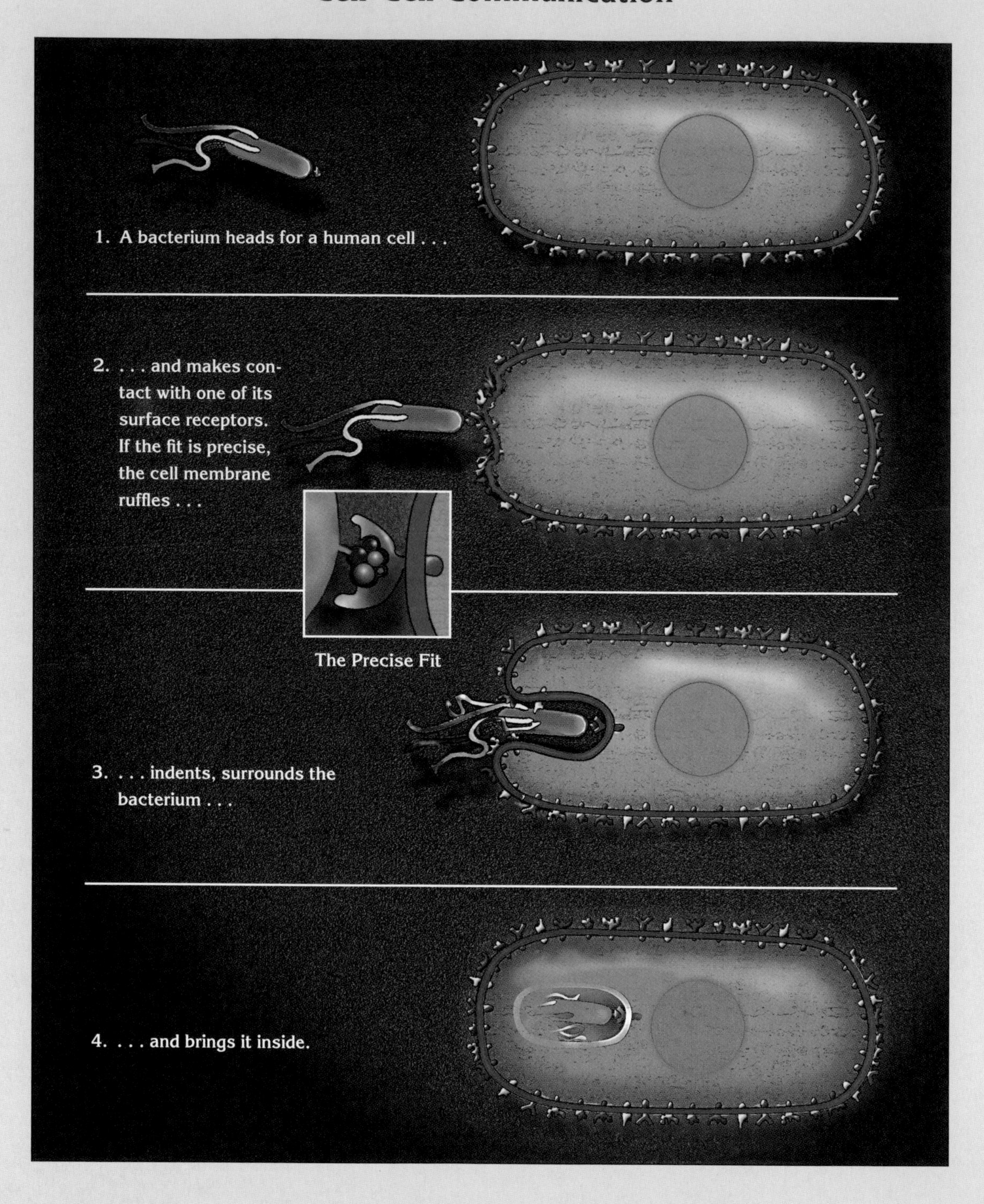

Certain microbes have evolved surface proteins—"keys"—that fit receptors—"locks"—on the surface of our cells. This lock-and-key fit initiates a false message like, "Hi, I'm food." Tricked in this way, our cells invite the invader inside. The bacterium's protein causes the cell membrane to ruffle outward to fit snugly around the bacterium. The invitation to come inside is fulfilled when the membrane sac surrounding the invader is drawn into the cell and gets pinched off. The bacterium then escapes from its vehicle and is free to multiply inside our cells and wreak havoc. We call such bacteria pathogens.

When Friends Turn Against Us

Our microbial partners join us from the moment we are born. They come from the first people we encounter—our mother, father, nurses and doctors, sisters and brothers. Over a matter of hours, we become home to huge numbers of microbial companions. They and we have begun an intricate partnership, an association that will continue for the rest of our lives.

Some of these microbial inhabitants remain steady partners. They stick to the various surfaces of our body, using specialized attachment sites that seem to have evolved just for the purpose. Others are passing through on their way to someone or somewhere else. Our microbial partners come and go as the years pass and as we change, but we will never be without a huge family of microbes.

Most of the time we live with these microbial inhabitants in a mutually beneficial relationship, exchanging valuable resources. Some of our steady partners make nutrients that we rely on for our health and well-being. Some produce substances that are noxious to other, less familiar microbes, protecting us from invasion. In exchange for their services, we supply them with a steady diet and a nice warm place to live.

Among the microbial members of our normal flora are some skilled opportunists, microbes just waiting for a chance to get an advantage. We can provide that chance through our own actions—enhancing a microbe's food supply, disrupting the normal microbial competition, or making a hole through skin or mucous membrane barriers—all enticing opportunities for members of our normal flora. Other microbes are less aggressive, but even a normally benign microbe can get out of hand, teasing the immune defenses of our border patrol into damaging our own tissue in the ensuing exchange.

Propionibacterium

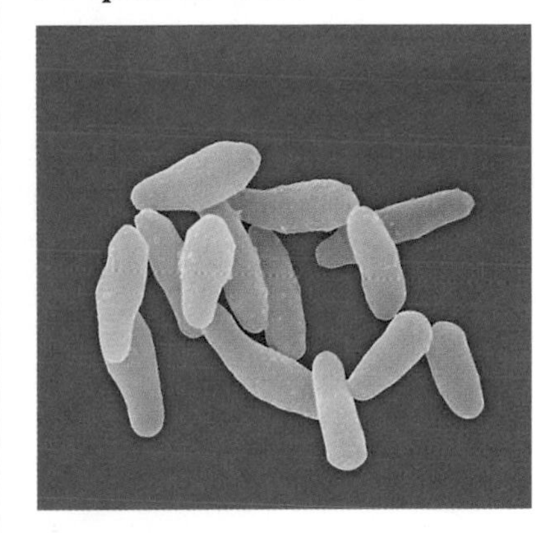

Identity: Bacterium
Residence: The pores on people's skin
Favorite pastime: Making noxious odors
Activities: A member of The Protectors, *Propionibacterium* creates compounds that protect us from other microbes.

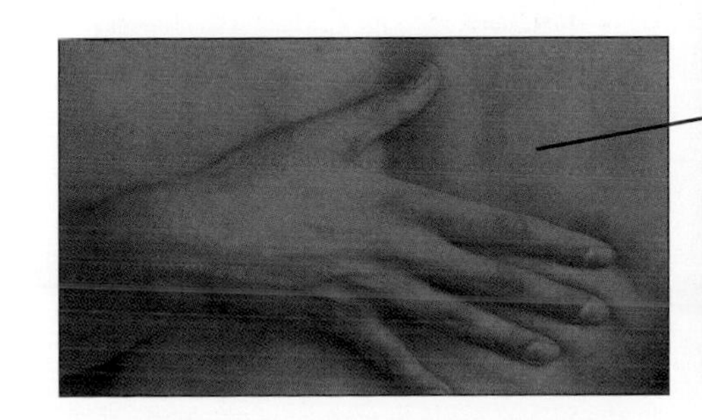

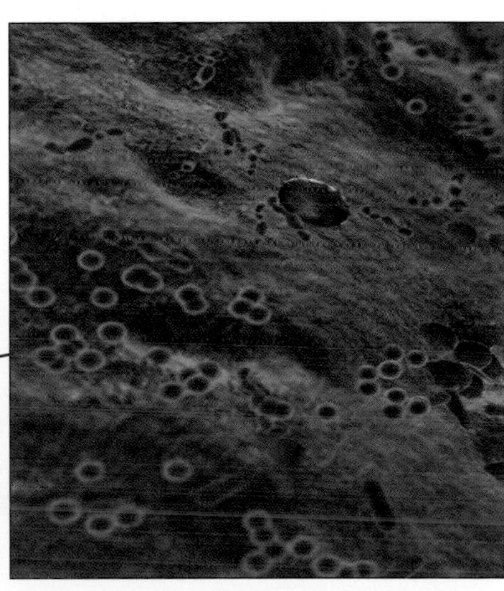

Our skin harbors at least thirty different kinds of microbes. On our backs there may be several hundreds to thousands per square centimeter; on damper surfaces, thousands of times that many.

Our Skin—Deserts and Rainforests

As adults, we have almost two square meters of skin. It creates an important barrier between everything else and us. From a microbe's view, our skin offers regions ranging all the way from the dry desert of our back to the tropical rainforests of our groin and armpits. Bacteria called *Staphylococcus, Corynebacterium,* and *Propionibacterium* attach to specialized receptors on our skin cells, each species preferring a different geography. Their densities vary like our own, from urban to suburban to rural. The highest densities are present around urban centers like our sweat glands and the sebaceous glands that lubricate our hair.

Staphylococcus epidermidis

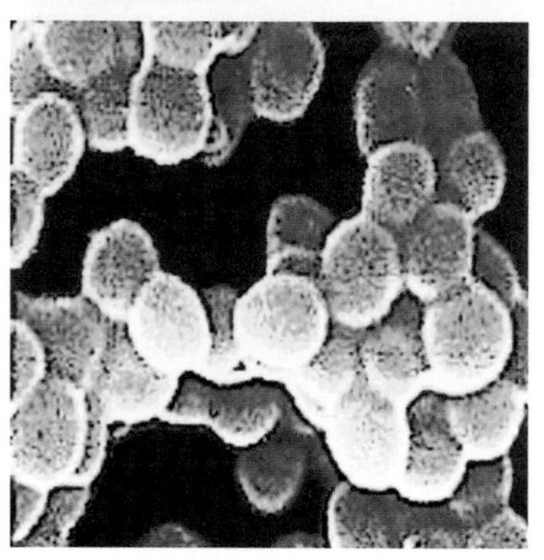

Identity: Bacterium
Residence: People's skin
Favorite pastime: Producing glue
Activities: A member of The Opportunists, *S. epidermidis* normally just occupies space on the skin and keeps harmful microbes away; slipping a catheter through the skin opens up opportunities for this microbe to cause disease.

Skin bacteria get much of their water and food from our sweat. In return, they produce noxious substances to guard their territory, repelling other microbes that—incidentally—may be less friendly to us. We know the skin bacteria are there because the substances they produce are the cause of body odor.

Knowing this, we might well ask, "Is washing a good idea?" We shouldn't change our perspective on the merits of bathing, however. Soap and hot water remove transients and help our microbial partners defend their turf.

The skin and mucous membranes are also our most important barrier to microbes. If we sustain a cut, we are exposing new territory to both our normal flora and potential pathogens (microbes that cause disease) that might be lurking about. Our immune system is quite familiar with members of our normal flora as well as frequent visitors when they appear in prohibited space. The killer cells of our border patrol can detect, surround, and dispatch these trespassers readily. Even when a small cut is infected, it usually heals quickly, leaving our normal flora outside where they belong.

Occasionally, we present opportunities to microbes in unexpected ways. Modern medicine, for example, relies heavily on a thin plastic tube called an intravenous catheter to put fluids and drugs directly into the bloodstream. The catheter penetrates through the skin, leaving an open pathway for microbial opportunists to grow between the catheter's outer surface and our tissues.

Some skin microbes, like *Staphylococcus epidermidis,* can attach themselves tightly to the catheter using a microbial glue they produce. Until the catheter is removed, they will enjoy their new habitat unencumbered by competition from other residents. Their vacation away from the skin surface, however, means disease for us. They cause wound infection around the catheter and can shed from its surface into the bloodstream.

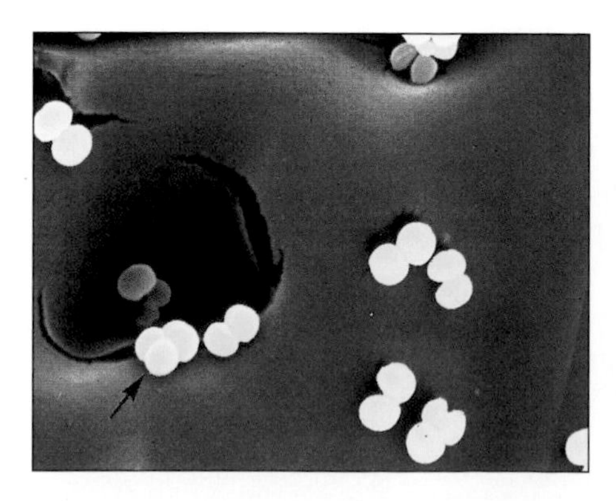

1. Microbes infecting this catheter are glued to its surface.

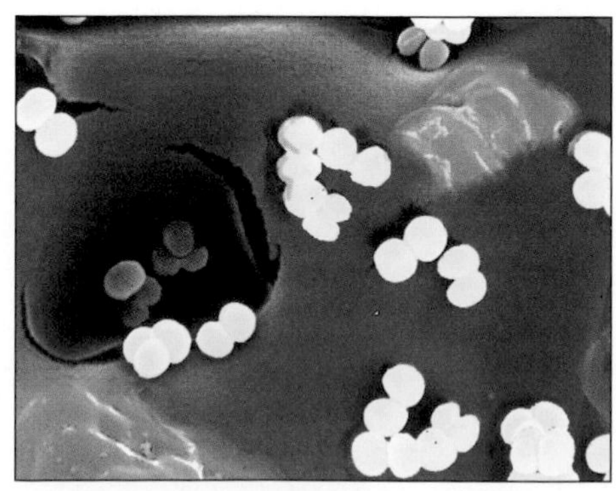

2. Once glued in place, they begin to divide.

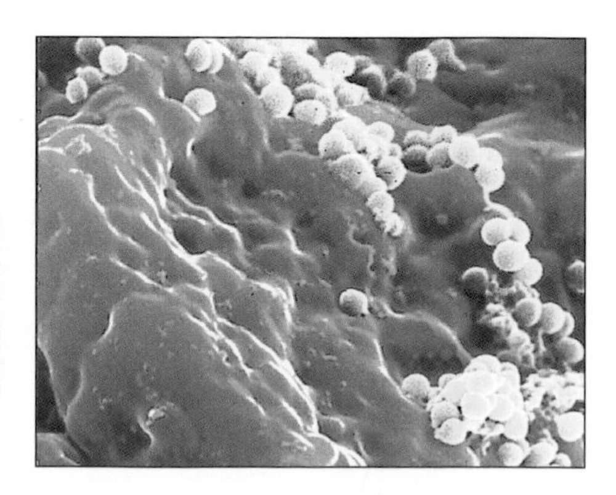

3. After enough time, the entire surface is coated with microbes and their glue.

The Mouth—Diversified Wetlands

Defined by surfaces like the teeth, tongue, and cheeks, the habitats of the mouth are bathed in nutrient-rich fluids along with digestive enzymes and saliva. At least 400 different microbial species live on the mouth's various surfaces. They live in a set of complex communities, competing with each other for space in some cases, and collaborating with each other for food and protection in other cases.

Some of the members of these communities come or go when we get our first teeth and some come or go when we lose them, but the competition and cooperation remain a constant feature of their existence.

The most common mouth microbes include bacteria named *Streptococcus, Lactobacillus,* and *Neisseria,* plus a vast array of bacteria called anaerobes. Anaerobes grow only in the absence of oxygen. The fact that our mouth is home to a large number of microbes that can't grow when oxygen is present may seem a bit unlikely. However, they attach to specialized sites deep in the crevices of teeth and gums, in miniature environments where oxygen doesn't penetrate.

The mouth microbes work much like those on the skin, taking up valuable space and helping prevent attachment of harmful species. You can recognize them by rubbing your tongue over your teeth after waking up in the morning. The material we recognize as plaque is really a dense microbial community growing in the protein layer on the tooth surfaces. Plaque is one version of biofilm, a covering blanket of proteins, sugars, and living cells, and "morning breath" is just another reminder of microbes at work.

Many of us learned growing up that eating too much sugar rots your teeth. This is true, but only indirectly. The real agent of tooth decay is a bacterium named *Streptococcus mutans.* It lives in about 90% of us, attaching to the crevices and fissures of our teeth with microbial glue called dextran polysaccharide. The microbe produces this glue from sucrose, or table sugar. The more sucrose we eat, the more glue *S. mutans* makes, and the more its foothold improves. Hanging on with its dextran glue, *S. mutans* makes acids and protein-destroying enzymes that demineralize the tooth surface and lead to tooth decay.

By halting this tenacious tooth microbe, we can prevent tooth decay. Fluoride in our drinking water helps to prevent tooth damage by halting the removal of the minerals that make the tooth surface hard. We can help avoid extra trips to the dentist by eating less sucrose and brushing our teeth regularly to remove the microbes and their glue.

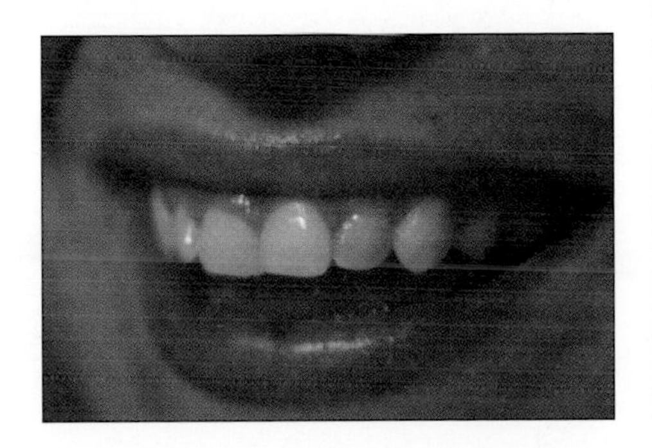

The mouth offers a hospitable environment for over 400 different kinds of microbes.

Porphyromonas

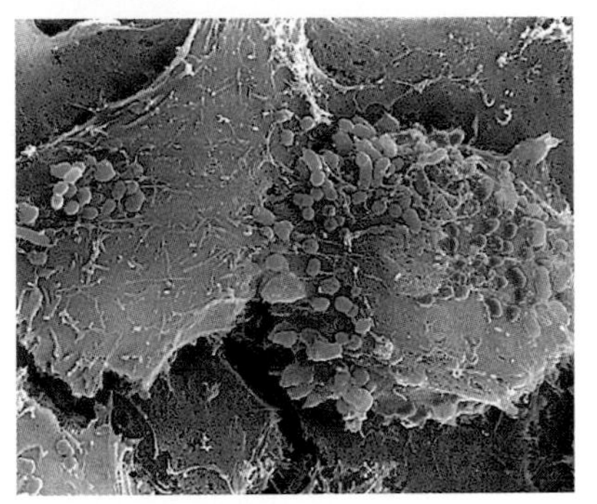

Identity: Bacterium
Residence: Gingival crevices in people's mouths
Favorite pastime: Promoting gingivitis
Activities: A member of The Opportunists, *Porphyromonas* lies in wait for an opportunity to grab space and produce a disease that causes people's teeth to fall out.

Lactobacillus

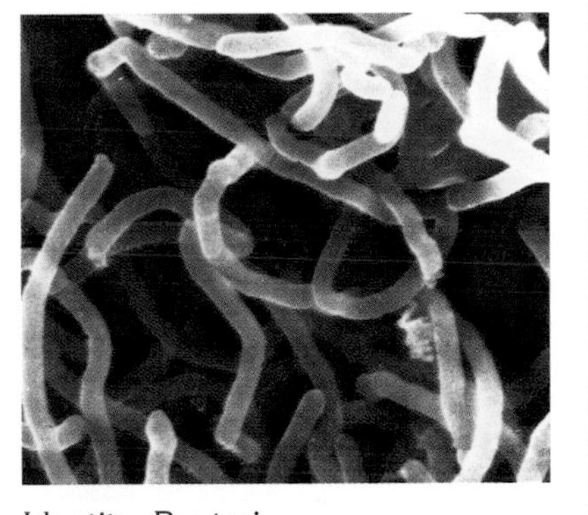

Identity: Bacterium
Residence: Certain mucous membranes of people
Favorite pastime: Making acid out of sugar
Activities: A member of The Protectors, *Lactobacillus* keeps conditions on the low side of pH 7.0, making the mucous membranes inhospitable to intruders; close relatives are involved in the yogurt business.

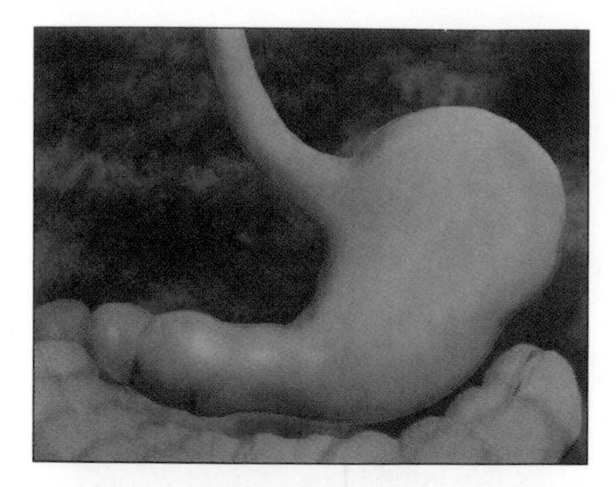

The stomach's acid environment is a powerful deterrent to microbes.

The Stomach and Intestines—Acid Lakes and Fermenting Bogs

Each of the three main parts of the digestive tract—the stomach, the small intestine, and the colon—is a unique environment. Each has a different role in breaking down the food we eat and conducting it to the cells inside our body. Each part also has its own unique microbial homesteaders.

The stomach is a scene of mass destruction for microbes. They are churned and tumbled in a highly acidic environment that only a few can endure. All that acid keeps certain disease-producing microbes in food and water—like *Salmonella*—from launching a successful invasion from the small or large intestine. We would normally have to eat a very large number of these microbes in order for a few of them to survive and cause disease.

For microbes, the small intestine is a treacherous bridge from the lethal chaos of the stomach to the relative utopia of the colon. Once in the small intestine, microbes must withstand detergent-like bile salts designed to rip microbes apart. They are twisted in slime and substances that act like meat tenderizers. Only the most hearty come through alive and enter the less chemically toxic world of the colon.

The small intestine, still hostile to those who have negotiated the stomach, offers some hope.

The large intestine, or colon, is a metropolis of microbial life. At least 500 to 600 different species live there as part of our normal flora. They begin colonizing the colon within the first few hours after we are born. Their member species increase and change as we vary the food we eat. They supply us with vitamin K, vitamin B_{12}, thiamin, and riboflavin, which we can absorb into our bodies. They maintain an intensely competitive and closed community, which makes invasion from disease-producing microbes a considerable challenge.

The colon is like a large fermentation vat, literally filled with microbes. In fact, almost half of our fecal material is made up of microbes. From day one, they are busily engaged, for their own benefit, in breaking up the food that we have eaten and partially digested. They convert the end products of our efforts into even smaller chemical compounds that we share with our resident microbes. When we eat corn, for example, we ingest a lot of

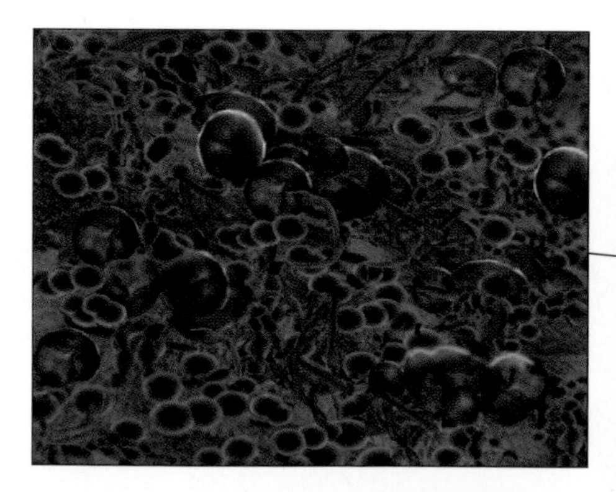

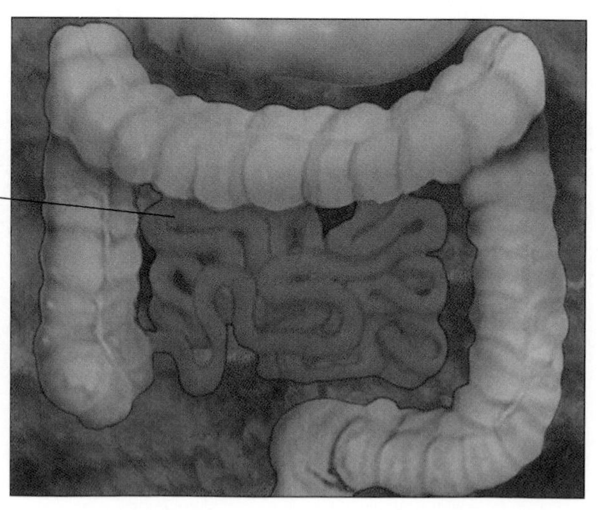

The colon is like a densely populated metropolis, an urban melting pot for microbes.

cellulose that we can't digest. One group of microbes produces digestive enzymes that break some of the cellulose into smaller chemical pieces, including glucose, which other microbes and we can use.

If we were taught that eating too much sugar rotted our teeth, we also learned that stress and stomach acid cause ulcers. Both contain a germ of truth. Most stomach ulcers, however, are a consequence of a mismatch between a bizarre microbe that lives in the stomach and our own immune system. This discovery is one of the more startling in the last decade, and allows physicians to cure most ulcers now with antibiotics.

The microbe, named *Helicobacter pylori,* burrows under the mucous coating of the stomach lining. *H. pylori* produces a substance that neutralizes the acid in its immediate surroundings, existing quite contentedly in an environment inhospitable to almost all other microbes.

By the time we enter midlife, up to half of us will be home to *H. pylori.* We provide our guest with space, water, and food, and yet we don't have ulcers. Why?

H. pylori is a microbe that teases. It provokes the immune system into a low-level but relentless attack that we call inflammation. Hence everyone who carries this organism has a symptom-free condition called gastritis—inflammation of the stomach lining. In a small number of people, something happens to intensify this relationship. The inflammation gets worse, and the sufferer ends up with plenty of symptoms and a stomach ulcer.

Without our normal intestinal flora, we do not thrive. In fact, if intestinal microbes are not present early in our lives, the surface of the intestine doesn't develop normally. The surface remains smooth rather than developing the carpet of projections called microvilli. Without microvilli, we would have to adjust to living with a vastly reduced ability to absorb water and nutrients.

Helicobacter pylori

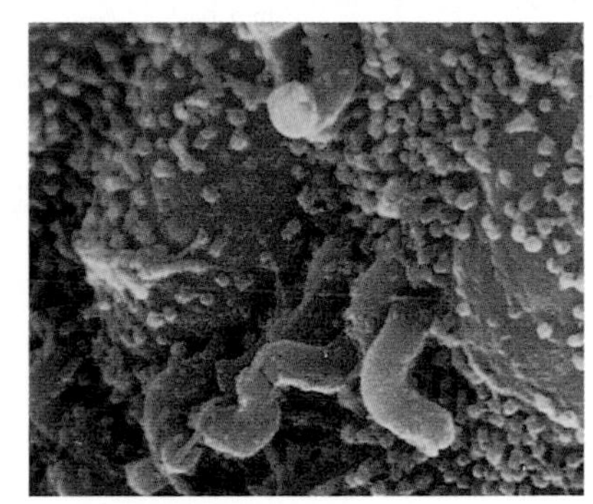

Identity: Bacterium
Residence: People's stomachs
Favorite pastime: Burrowing under the mucous coating that covers the stomach surface
Activities: Another member of The Opportunists, *H. pylori* sometimes creates conditions in the stomach that lead to ulcers.

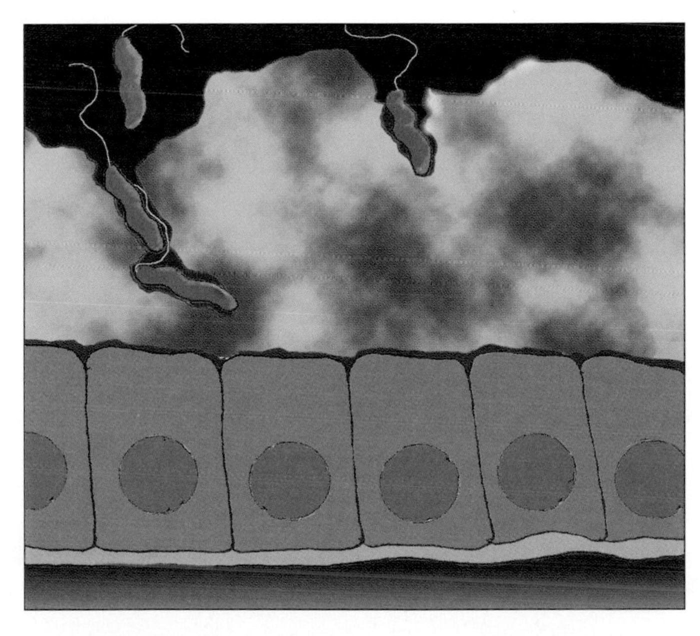

H. pylori cells dive into the thick mucus coating the stomach lining . . .

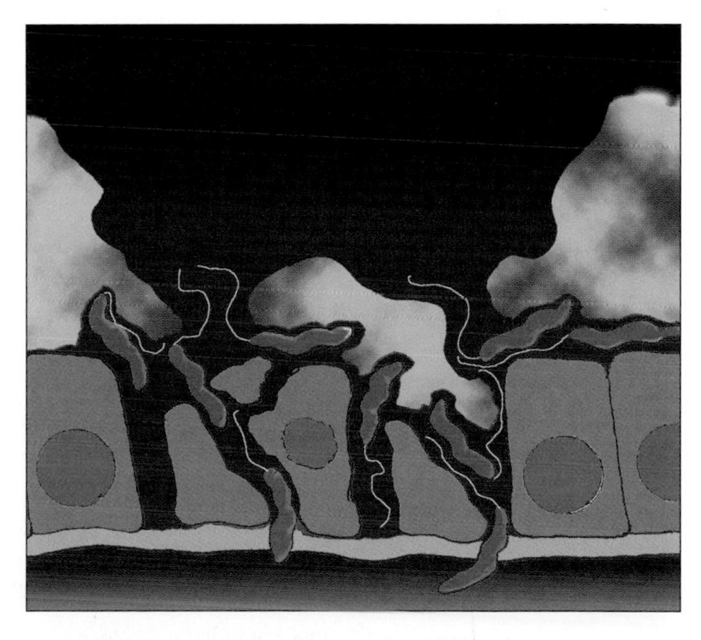

. . . where their toxins, combined with the acid environment and our own defensive responses, diminish the mucus layer and lead to damage.

Escherichia coli

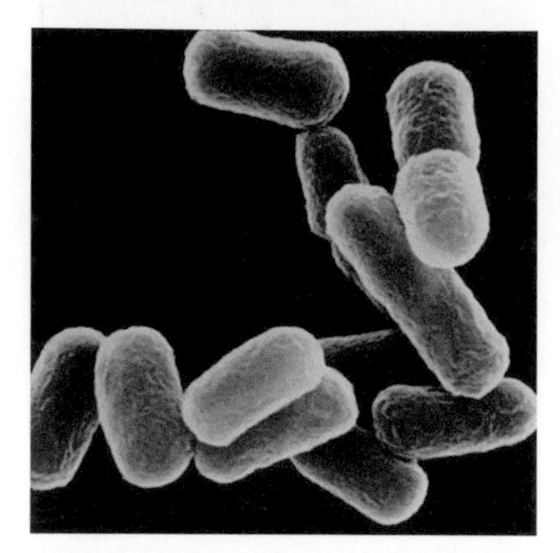

Identity: Bacterium
Residence: The large intestine of people and other warm-blooded animals
Favorite pastime: Fermenting
Activities: A member of The Protectors, *E. coli* uses up the excess oxygen in the intestine, making conditions right for other bacterial members of the intestinal community to thrive; when healthy, this community prevents incursion of disease-producing microbes.

As important as our intestinal microbes are, we may not be inclined to appreciate all of their activity. One unavoidable feature of microbial metabolism is gas: hydrogen, carbon dioxide, methane, and other odorous compounds. We experience their activity as flatulence.

Antibiotics have helped to save many lives since their advent five decades ago. But most are not very specific or selective. In other words, an antibiotic will kill the members of our normal flora along with an invading microbe. Those remaining are more than happy to take advantage of the lull in the competition to expand their territory.

Clostridium difficile is just such an opportunist. If this microbe is present in our colon, its population is usually quite small. However, it survives the killing action of many antibiotics when numerous other species don't.

When some of the competition is removed, *C. difficile* must experience something like microbial nirvana—plenty of food and plenty of space to reproduce in the absence of normal competition. And reproduce it does. Its population can grow into the millions in the space of only a few hours.

C. difficile shows no gratitude for getting this momentary advantage. As its population expands, the microbes produce toxins, which cause destruction and death to cells that line the small intestine. The end result is diarrhea—and sometimes not just a benign upset. In some people, the disease is severe enough to require hospitalization; in a few, it is fatal.

C. difficile and its toxins are just one of many reasons to avoid taking antibiotics when they are not effective or necessary.

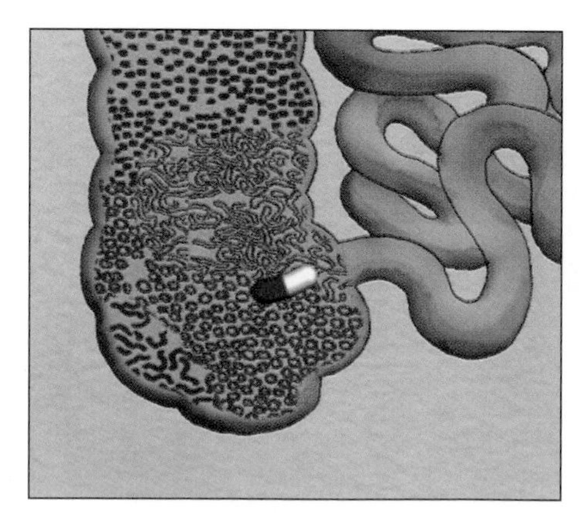

1. Bacteria in the colon . . .

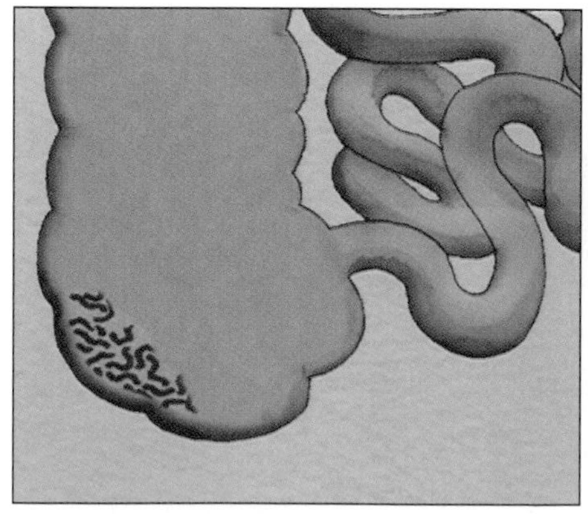

2. . . . are killed off by antibiotics . . .

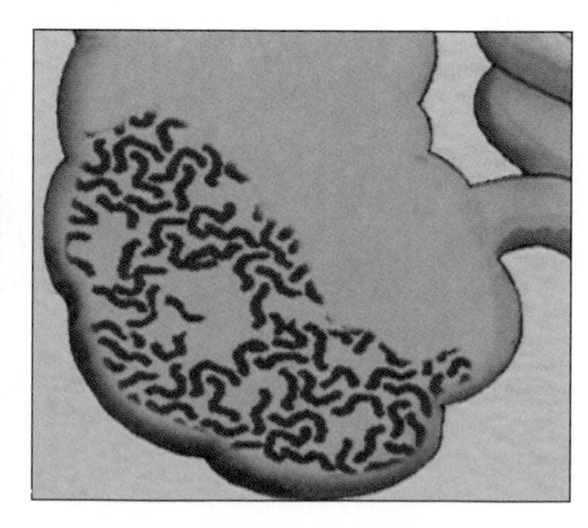

3. . . . leaving the field open to opportunists like *C. difficile.*

A Bold Experiment

Barry Marshall, a young Australian physician, stood in his laboratory and prepared to drink the cloudy fluid in his glass flask. This was a strange thing for him to do since he was fully aware that the fluid was teeming with a small, comma-shaped bacterium called *Helicobacter pylori,* a microbe he was convinced caused stomach ulcers. Saying, "Here goes," he raised the flask to his lips and drained it. In so doing, Marshall was using one of the earliest traditions of medical research—infecting himself to prove that the microbe he had been studying was indeed the culprit in causing disease.

At the time, his theory flew in the face of conventional wisdom. Physicians had been trained for decades that ulcers were a consequence of a psychosomatic event—stress. This was such a firmly held view that Marshall and his mentor J. R. Warren were having considerable difficulty getting their scientific colleagues to take their theory seriously. So Marshall set out to demonstrate in this very personal way that their arguments were credible.

To prove that a given microbe causes a specific infectious disease, scientists use a set of criteria called Koch's postulates. Marshall and Warren had proven two of these criteria—that the little microbe was associated with stomach ulcers and that it could be taken from the stomachs of ulcer sufferers and grown in the laboratory. Proving the third criterion was to be much more difficult—demonstrating that the microbe would cause ulcers when it infected a new host. Marshall had reached a dead end in his search for a suitable animal model for the new host. Then it occurred to him to simply use himself.

In preparing for the event, Marshall had to undergo examination of his own stomach lining to make certain it was healthy. This alone might have deterred a less dedicated person, since it meant passing a tubelike device called an endoscope through his mouth and into his stomach. Satisfied that he was uninfected and undamaged, he proceeded to infect himself.

After drinking the live culture, his stomach began to growl, and within a week he was feeling nauseated, tired, and unusually hungry. Marshall returned to the physician who had examined his stomach lining so that he could collect new specimens. Sure enough, the microbe was growing happily away in a red and inflamed area of his stomach, thus reproducing some of the symptoms and physical findings associated with ulcers. He had fulfilled the third criterion.

When asked to comment on the occasion, Marshall merely noted that the culture tasted like swamp water.

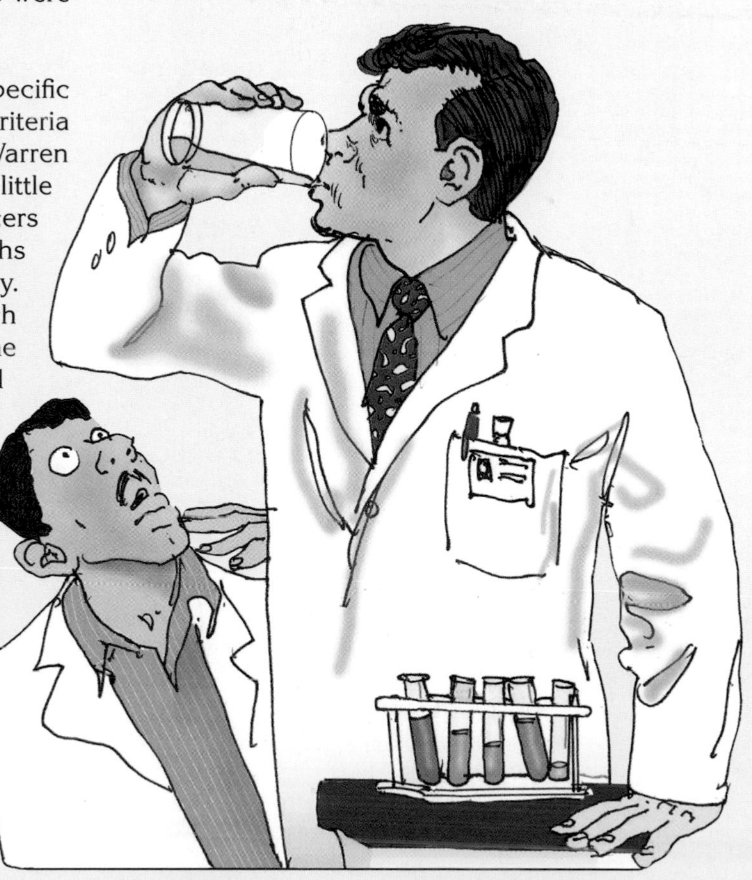

Familiar Enemies

Species that are closely intertwined tend to evolve toward cooperation. Over time, even putative enemies tend to settle into some form of accommodation. They become so familiar that enemies almost appear as friends.

An Enemy That Is Usually Just Passing Through

Consider *Salmonella typhi*. This bacterium causes typhoid fever, a disease that is very different from the intestinal upset we get from other *Salmonella* species. After thousands of years of contact with us, *S. typhi* has come to depend on us for survival. It doesn't infect other animals. Across the millennia, the bacterium has learned enough about us to be a skilled invader, sending misleading chemical messages to our cells and gaining easy entrance. We, in turn, have learned something about mitigating the bacterium's lethal properties.

We become infected with *S. typhi* by eating contaminated food or drinking contaminated water. Once *S. typhi* has survived its journey to the small intestine, the bacterium uses its small whiplike flagella to swim close to the intestinal wall, all the while monitoring its environment. When conditions of pH and oxygen concentration are just right, it pushes a protein through its outer surface. The protein becomes its docking device. Along the inside surface of the small intestine are specialized cells called M cells. They have sites on their surface that the bacterium's docking protein just fits. As the bacterium touches down and locks on, it signals to the cell, perhaps saying, "Hi. I'm food. It's dinner time!" The M cell, misled by the greeting, responds by ruffling its outer membrane, reaching out, and drawing the bacterium inside. Once inside the cell, *S. typhi* begins to reproduce rapidly, free from the pressures and competition in the intestine.

From here, the plot thickens. The newly minted bacteria break free from the M cell and invade deeper into the tissues surrounding our small intestine. They send out a new message, saying, "I've slipped past your defenses, so come and get me if you can."

S. typhi's taunt reaches the cells that patrol our tissue for invaders. We call these cells—part of our immune defenses—macrophages. Macrophage means "big eater," and the name is apt. Macrophages scout constantly, picking up foreign material and microbes that have wandered into the wrong spots in the body. Normally, they ingest intruders, subjecting them to a barrage of toxic chemicals and bringing them to a horrible death.

The *Salmonella* Journey

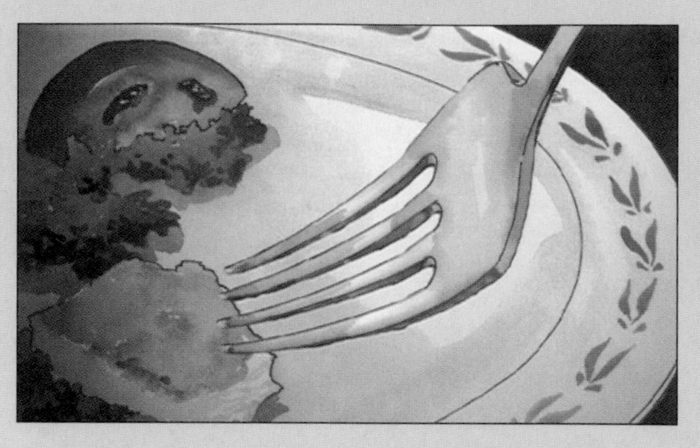

1 *S. typhi* usually begins its trip through us when we eat contaminated food or water.

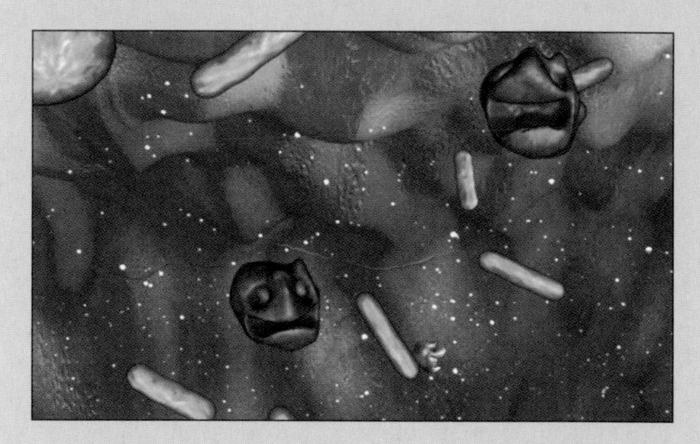

4 After that acid test it faces an onslaught of bile and other chemicals in the upper small intestine.

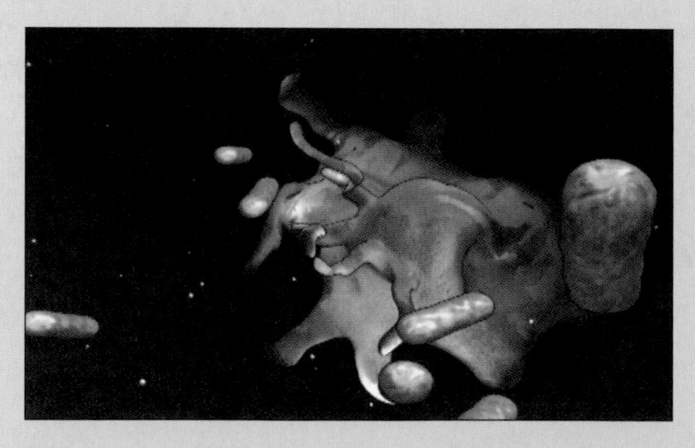

7 Here the bacteria encounter macrophages roving among the cells surrounding our intestine and are gobbled up.

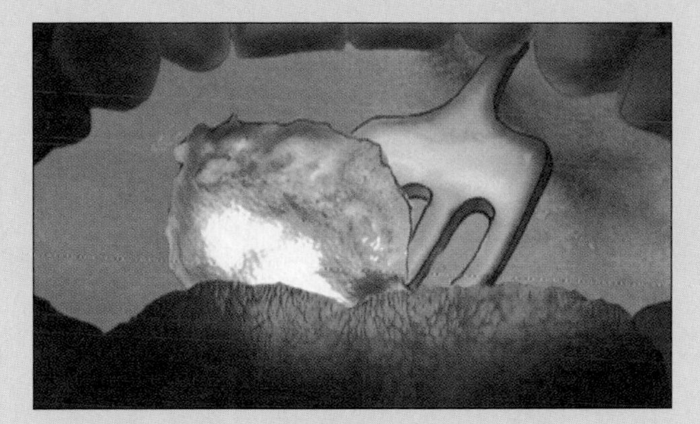

2 The bacteria survive our first line of defense—digestive enzymes in the saliva.

3 *S. typhi* must still struggle through the extreme acid conditions in our stomach.

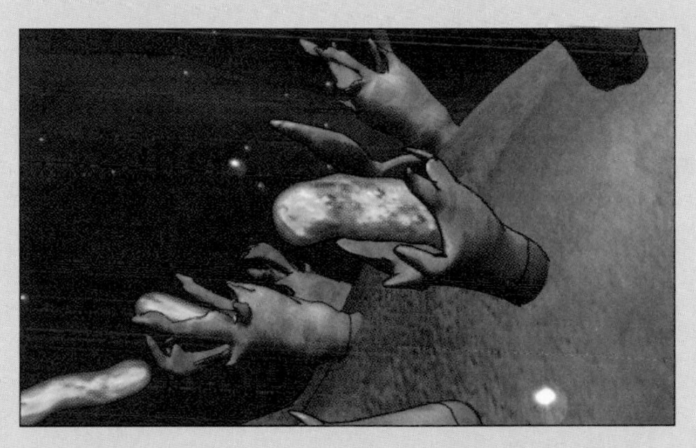

5 *S. typhi* now moves quickly from "in side" to inside, entering specialized cells, called M cells, that are part of the wall of the intestine.

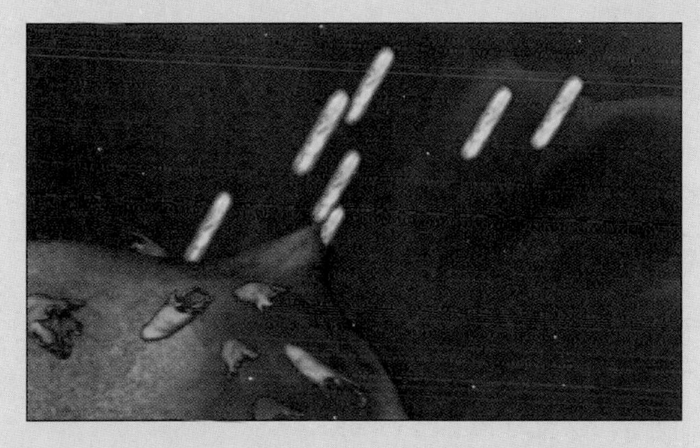

6 The invading bacteria multiply inside the M cells, then penetrate the tissue adjacent to those cells. Now they're REALLY inside.

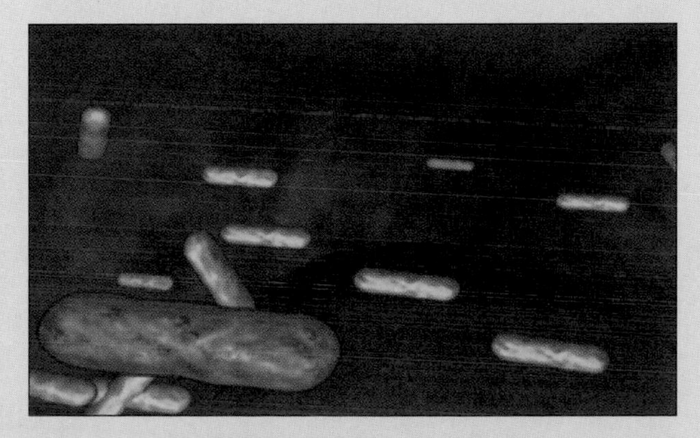

8 The macrophages would normally kill bacteria, but *S. typhi* chemically stuns the macrophage's killing device and continues to multiply.

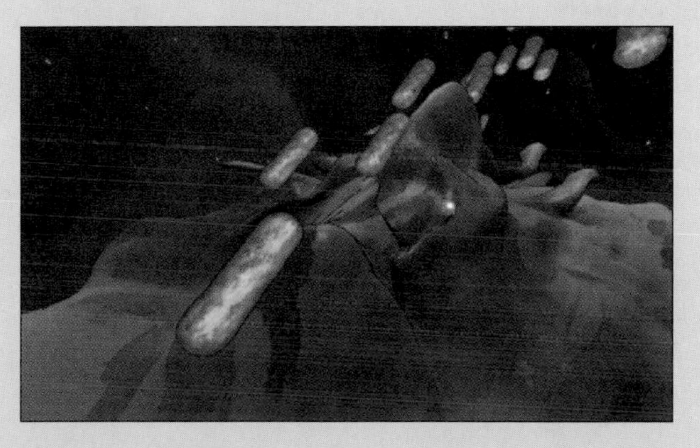

9 The macrophages return to the bloodstream, where the bacteria break out to infect other cells, and we get typhoid fever!

Not so for *S. typhi*. Having "heard" the chemical taunt, the macrophage reaches out and gobbles up the bacteria. *S. typhi* now changes its tactics, convincing the macrophage to forego its normal malevolent behavior. The bacterium not only survives but also thrives, safely reproducing yet again inside the macrophage.

The infected macrophages return to the bloodstream, carrying *S. typhi* along with them. From there the bacteria can invade new macrophages in the spleen and liver, getting food and reproducing all the while.

S. typhi causes us considerable discomfort, but our immune system—or the antibiotic we take—usually kills the bacteria. By then, however, the bacteria have successfully reproduced, some going on to infect other susceptible people. This is typical of an encounter with the most familiar of enemies—the result of messages sent between two old adversaries who have co-evolved and learned to coexist.

Typhoid Mary

Salmonella typhi may only be a familiar microbe to anyone who has recently traveled into a developing country. But back at the beginning of the 20th century, you didn't need to travel to find it. You could get it with your meal at the local restaurant.

The now immortalized "Typhoid Mary" Mallon was a professional cook in New York City at the turn of the century. After several outbreaks of typhoid fever in the City, the local health officials, operating in the fine tradition of John Snow, the first infectious disease detective, traced the outbreaks to Mary, who was ultimately identified as a chronic carrier of *S. typhi*. She had been serving up the microbe along with her cuisine.

She was offered a solution. The microbe lives in the gall bladder in chronic carriers, and removal of the infected gall bladder was the only remedy known. Mary refused the operation and refused to cease cooking, at which point the constabulary promptly arrested and incarcerated her. After three long years, Mary agreed never to cook professionally again, and she was released.

But Mary apparently loved to cook. She immediately changed her name and resumed her profession. Typhoid Mary was employed by a variety of hotels, restaurants, and hospitals, from which she managed to spread *S. typhi* to many more people before she was finally tracked down again. She spent her remaining years quarantined in a New York City hospital.

Familiar Enemies Everywhere

We are host to a reasonably large number of disease-producing microbes with which we have apparently coevolved. Like *S. typhi,* they and we have seemingly reached some level of accommodation.

Some regularly cause disease, but rarely kill. *Streptococcus pyogenes,* the microbe that causes strep throat, accounts for over 30 million cases of disease annually just in the United States, yet few people actually die from an infection with this microbe.

Others regularly inhabit us but rarely cause disease. *Neisseria meningitidis* resides benignly in up to 10% of people's throats, yet it causes less than 2,000 cases of a disease called meningitis every year in the United States.

Our familiar enemies include *Streptococcus pneumoniae,* the microbe that causes a common form of pneumonia; *Neisseria gonorrhoeae,* a microbe that causes the sexually transmitted disease gonorrhea; and *Staphylococcus aureus,* one of the most common causes of skin and wound infection. Although each of these microbes can and does produce disease in us, each can, and frequently does, live in a state of détente with us.

Group A *Streptococcus*

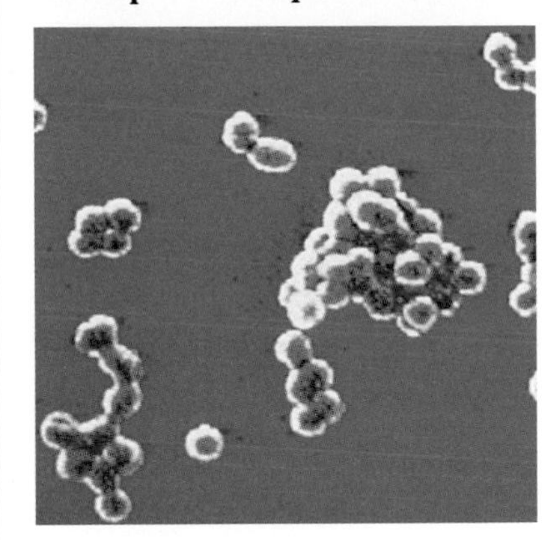

Identity: Bacterium
Residence: People's throats
Favorite pastime: Jumping from kid to kid
Activities: A member of the Human Pathogens, group A *Streptococcus* causes "strep" throat; rarely, it causes very serious disease, invading tissue and earning the popular moniker "flesh-eating bacteria."

Infection Doesn't Necessarily Mean Disease

While ten people out of a population of 100 may be infected by one of our familiar enemies, only a few of these people will actually suffer damage brought about through the microbe's interactions with them. Scientists call the former colonization and the latter infectious disease.

Coevolution Leads to Accommodation

Scientists believe that part of the reason we seem to have such a wide spectrum of interactions with microbes is that we are looking at diseases at different stages of evolutionary development. At one extreme, there's a deadly but dead-end relationship, like that between the Ebola virus and us. At the other extreme, there's a mostly beneficial relationship, like that between the members of our normal flora and us. In between, there's a far less deadly and more familiar relationship, like that of us with *S. typhi* or pathogenic *E. coli.*

If we could look at our interactions with a specific microbe, one that is evolving along with us, over a very long period of time, we might be able to witness a change from a lethal pathogen like the Ebola virus to a more moderate pathogen like *S. typhi.* Maybe, ultimately, we would see such extensive accommodation between us that the microbe would cease to be a threat.

Although such observations are almost never possible within a single human life span, scientists have identified a few short-term events that give us clues that coevolution may lead to some degree of accommodation. Such is the case with the farmers, rabbits, and viruses in Australia.

The European rabbit was intentionally introduced into Australia in the 19th century. What seemed like a good idea at the time turned out to be something of an ecological disaster. The rabbit had no natural predators to control population growth. Rabbits did what rabbits do—like bacteria, they ate and reproduced. Before long, there were so many rabbits that they became a serious pest, destroying the Australian farmers' crops along with the local vegetation.

Scientists observed that a certain kind of virus carried by Brazilian rabbits caused lethal epidemics in European rabbits. They attempted to turn this observation into a practical solution to the growing rabbit problem in Australia by intentionally infecting the Australian rabbits. They were hoping to use the virus to impose a natural biologic control. The strategy was

The rabbits

Most rabbits were highly susceptible to the virus. These were killed off—until the less susceptible (shown here in brown) reproduced and became the majority population.

wildly successful during the first year, at least from the farmers' perspective. The virus killed almost 100% of the rabbits that it infected, substantially reducing the rabbit population.

This represented a windfall for the virus and a partial relief for the farmers over the short term. However, it was not in the long-term evolutionary interest of either the virus or the rabbit. If the virus killed all the rabbits, the way Ebola kills people, then it would reach a dead end. There would be no more rabbits that it could infect, and it would become extinct. If the rabbits were reduced in number to the point where they could not find mates, they too would become extinct. Although this might be good for the farmer, it would definitely not be good for the rabbit or the virus.

A few rabbits did survive, however, and still did what rabbits do—eat and reproduce. The next year, the virus killed only 90% of the infected rabbits, and with each successive year, the death rate among the rabbits fell, eventually dropping to 25%. What had happened?

Most of the rabbits that survived had natural resistance to the worst effects of the virus. They passed the genes conferring resistance to their offspring. With each succeeding generation, disease-prone rabbits were eliminated and resistant rabbits thrived. At the same time, the virus evolved. The most successful viruses caused a disease that was less rapidly fatal. They were consequently more likely to be passed to a new rabbit. The rabbits infected by the less lethal version could wander around for a considerable time while infected, transmitting their virus to lots more rabbits. Consequently, the population of less lethal virus progeny increased and ultimately became dominant.

Natural selection was acting on both rabbit and virus, each accommodating to the other. The virus and the rabbit live on to this day in Australia. And the farmers have yet to solve the problem of the rabbits eating their crops.

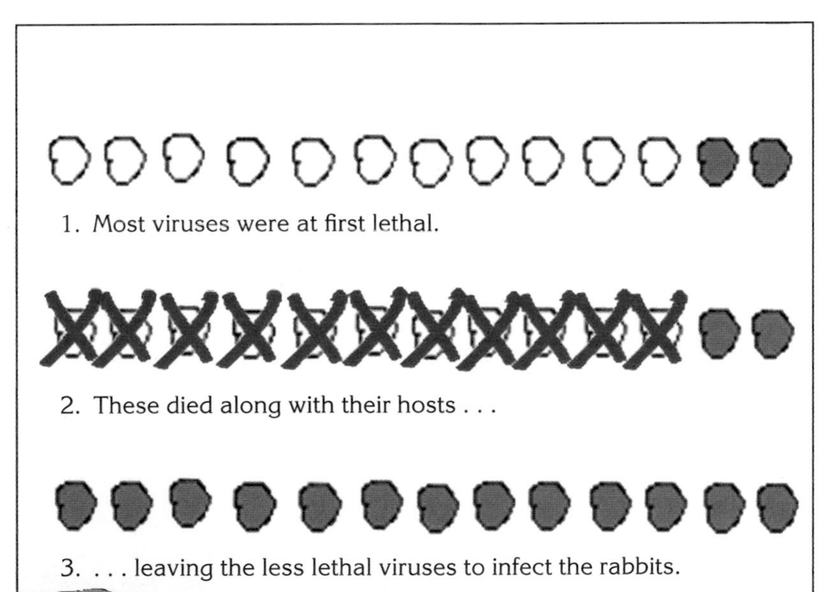

The myxoviruses
Most viruses were at first lethal. As they killed their host rabbits, they, too, reached a dead end. Their less lethal forms (shown here in blue) became the majority population.

Sickle Cell—the Lesser of Two Evils

Malaria, a disease affecting 200 to 300 million people annually, has afflicted humans since before recorded history. It offers a striking example of accommodation between microbes and people.

One of the microbes that cause malaria, named *Plasmodium falciparum,* causes a severe and potentially fatal disease, killing as many as a million children a year in certain areas of the world. Many of the people who live in these areas have a genetic disease called sickle cell. The genetic defect causes an abnormal form of hemoglobin to be made in the victims' red blood cells.

The abnormal hemoglobin differs by only one amino acid from the normal version, but that tiny change alters the shape of the molecule. When the oxygen concentration inside an affected red blood cell is low, the abnormal hemoglobin becomes rod shaped and links with other similar molecules of the abnormal hemoglobin, crystallizing inside the red blood cell.

Plasmodium

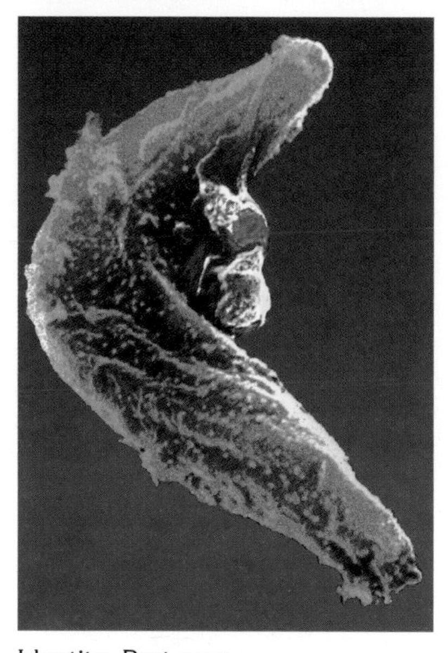

Identity: Protozoan
Residence: Mosquito guts and people
Favorite pastime: Circulating in red blood cells
Activities: A member of The Human Pathogens, *Plasmodium* causes malaria, a disease that has taken its toll on humanity, in some cases rewriting history.

The affected cells are rigid and shaped like a sickle, while normal cells are flexible and shaped like a donut. The normal mechanisms of the body remove the sickle cells more quickly than the normal cells, which can lead to severe anemia. Sickle cell disease is potentially fatal, so there should be strong selective pressure to eliminate the genetic defect in the population. Yet the disease is present in up to a quarter of the people from these areas.

Why should this be true? The microbes that cause malaria spend one essential stage of their life cycle infecting red blood cells. Scientists have learned that when the microbe infects the red blood cells of a victim with sickle cell, it causes even more cells to sickle, probably by reducing the oxygen concentration inside the red blood cell. As a result, the body removes the infected cells more rapidly than cells in an unaffected person. Because many parasites are removed and destroyed with the infected cells, falciparum malaria is less likely to be fatal. Consequently, children with sickle cell hemoglobin are more likely to survive a malaria infection and pass the gene for the defective hemoglobin on to their children.

Living with falciparum malaria has been such a high evolutionary priority that even a seemingly adverse genetic defect, like sickle cell disease, has been favored to survive.

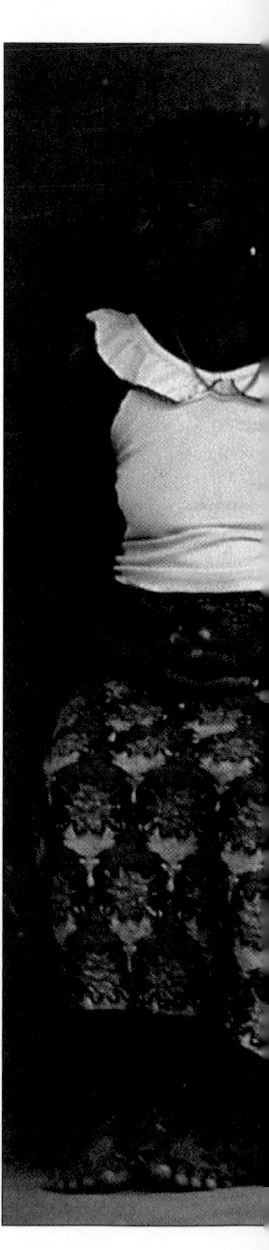

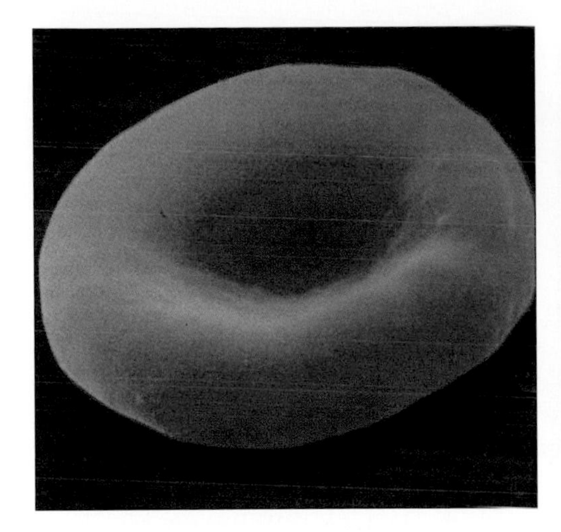

1. The normal red blood cell is a smooth, rounded, flexible disk that slips easily through the narrowest blood vessels.

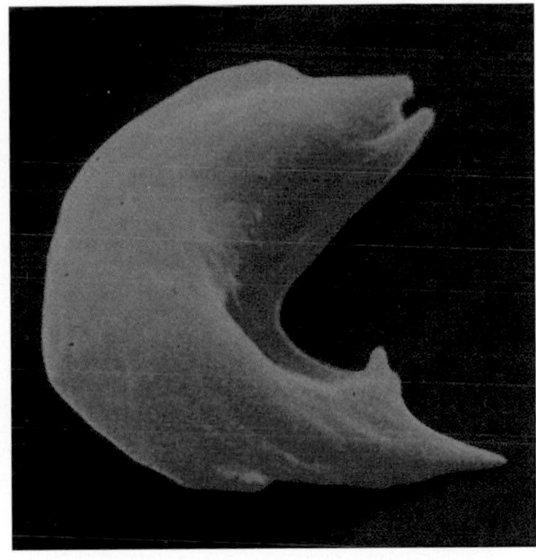

2. In sickle cell disease, the cells become elongated, shaped somewhat like a sickle, and more fragile. The sickled cells are removed quickly from circulation.

The Most Lethal of Strangers

Some infectious diseases appear suddenly or dramatically increase in a population within a short period of time. Scientists refer to this phenomenon as emergence. Medical researchers have identified at least twenty emerging infectious diseases in the past two decades. Some, like toxic shock syndrome and Lyme disease, are a consequence of new products or expanding our habitat. Others are a consequence of evolutionary changes in a specific pathogen, giving it new mechanisms to bypass our immune response (like the Hong Kong influenza virus) or resistance to our antibiotics. Still others, like the hantavirus pulmonary syndrome, are a consequence of a series of environmental factors that offer new opportunities for microbial expansion. Most are among our most dangerous of enemies.

Ebola virus

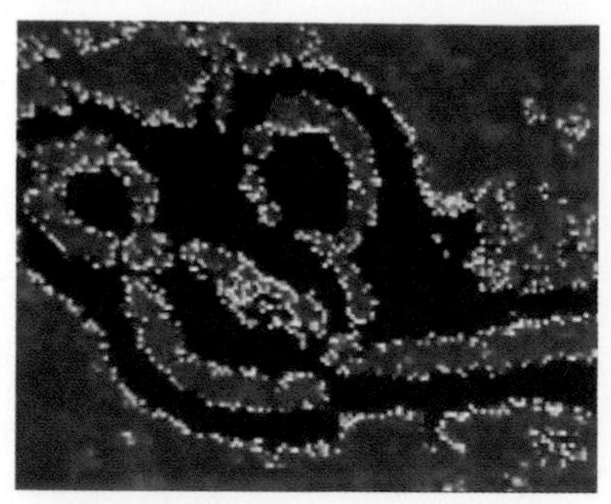

Identity: Virus
Residence: Somewhere in Africa, but no one really knows
Favorite pastime: Still a mystery
Activities: A member of The Lethal Agents, Ebola has a life cycle that doesn't usually include humans, but when it jumps from its still unknown hosts, it can cause deadly outbreaks of disease in people; the virus starred opposite Dustin Hoffman in "Outbreak."

Brief Encounters of the Worst Kind

The microbes responsible for some emerging diseases normally live with other creatures and infect us by accident. Such microbes may not be particularly harmful to their normal hosts, but they are strangers to us—and often our most lethal enemies.

Legionella pneumophila, the bacterium that causes Legionnaires' disease, normally lives with amoebae in fresh water. It infects us when we breathe aerosol droplets of water contaminated with the organism, causing outbreaks like the one at the 1976 American Legion Convention in Philadelphia, where 34 died and 221 were hospitalized. Scientists found the microbe and its companion amoebae living in the water in one of the hotel's air-conditioning cooling towers.

Legionella pneumophila

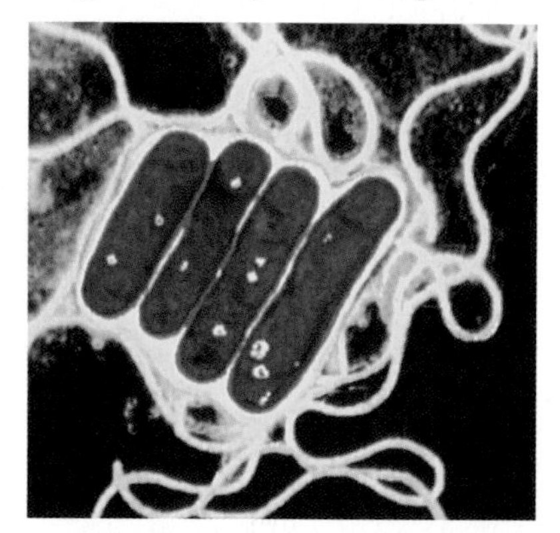

Identity: Bacterium
Residence: Lives with free-living amoebae in streams and lakes all over the world
Favorite pastime: Hanging out in hot water tanks and air-conditioning cooling towers
Activities: A member of The Opportunists, *Legionella* causes pneumonia if inhaled in large enough numbers.

Even more dramatic are the outbreaks of viral diseases like the mysterious Ebola in northern Zaire and southern Sudan and the hantavirus in the southwestern United States. From their yet-undiscovered hosts, the viruses spread among people in small, isolated populations, rapidly killing their victims within a few hours to a few days.

Although threatening to us, these outbreaks are often not really successful from the microbes' perspective. The bacteria that cause Legionnaires' disease cannot be transmitted directly from person to person, and so by infecting people, the microbes reach a reproductive dead end.

The Ebola virus can be passed through close contact from person to person, but often kills its victims so rapidly that it reaches an evolutionary dead end as well. There are no people left to infect. In stark contrast to our "familiar enemies," like *S. typhi,* this microbe seemingly comes from nowhere, swiftly invades, wreaks devastation, and then disappears.

The Centers for Disease Control and Prevention laboratorians work with highly pathogenic microbes in a high-level containment facility.

The Wrong Host at the Wrong Time

Dogs, cats, skunks, raccoons, and bats can all harbor rabies virus, which causes a deadly infection in them. We encounter the virus, which causes the same fatal disease in humans, if we are bitten by a rabid animal.

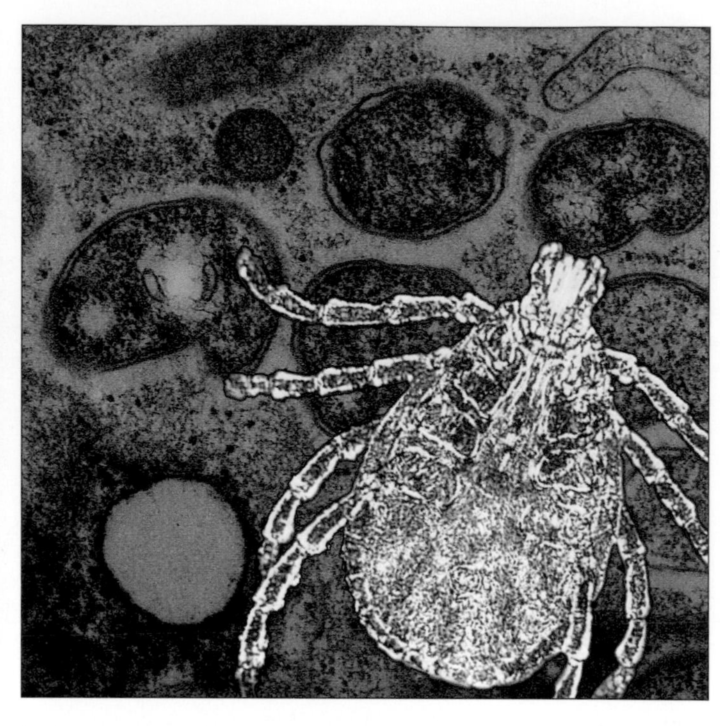

Ticks transmit a bacterium named *Rickettsia rickettsii* among its normal rodent hosts. We get Rocky Mountain spotted fever when we become a meal for the tick, which transmits the bacterium to us.

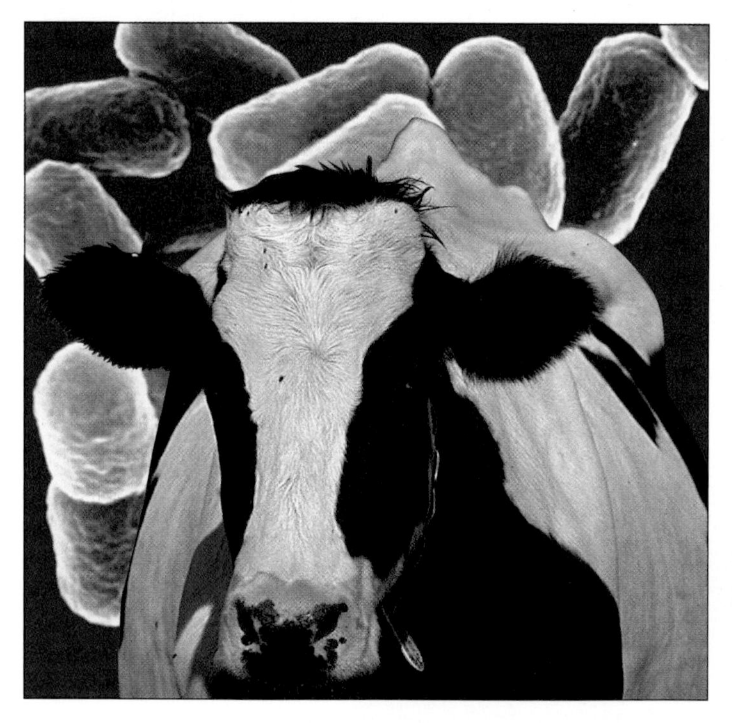

Cattle are a common host to a bacterium named *Brucella abortus,* which causes the disease infectious abortion. We get infected with *Brucella* through contact with meat and unpasteurized dairy products from a diseased cow.

Parrots and other birds are subject to an epidemic disease caused by a bacterium named *Chlamydia psittaci*. The bacterium, which causes pneumonia in us, can be transmitted to humans when we come in contact with an infected bird.

Microbes That Changed History

"And I . . . with my own hands, buried five of my children in a single trench, and many others did the like . . . And no bells rang and nobody wept no matter what his loss, because almost everyone expected death . . . And people said and believed, 'This is the end of the world.'"—Siennese chronicler, 1354 (Schebill, F., "Sienna," in *The Story of Civilization,* Part V, *The Renaissance,* by Will Durant.)

Emerging infections are not a new phenomenon. Some have changed the course of history.

Plague, or the Black Death, appeared and became epidemic in the cities in western Europe during the Middle Ages. *Yersinia pestis,* the bacterium responsible for the deadly disease, is transmitted among rats and other rodents by infected fleas. Fleas infesting the brown rats that arrived on the trade ships from the Mideast probably carried the plague bacterium along with them. The native English rat was resistant to the fleas, but the brown rat was more aggressive and quickly outcompeted the locals, thus establishing a new reservoir for the plague bacterium.

People were infected when the hungry fleas feasted on them rather than rats. This provided a great opportunity for the bacteria to expand into a new host, which they did with unbridled enthusiasm. Once infected, some of the people spread the offenders directly from person to person through the air, thereby eliminating the need for the flea intermediary. This was so efficient that over a period of four years, about a quarter of the population in western Europe died from the disease.

The Black Death killed so many people that western European society took a full century to recover. Laborers were in such short supply that wage labor became increasingly attractive. This allowed many people to escape from the feudal restrictions of the time and likely contributed to the demise of the feudal system in general.

The ability to spread from person to person is a significant advantage for any microbe if it is to survive in a human host. In the case of plague in the Middle Ages, the microbe began infecting people when they came into contact with rats carrying the microbe's host—fleas. The organism's success in finding a new host was magnified when it began to spread directly from person to person. Since people were living in close quarters and in crowded cities by then, the plague bacterium found a grand new opportunity for biologic success.

Other infectious agents have left their mark on civilizations as well. The smallpox virus, introduced into the Americas by Spanish explorers, accounted for the fall of two major empires. Smallpox killed millions of members of both the Aztec and the Incan empires, including their emperors. This made possible the later conquest by the small armies of Cortez and Pizarro. The English sweating sickness, a mysterious disease that haunted Tudor England from 1485 to 1551, had a similar impact on the population. The disease took the life of Prince Arthur, thus paving the way for his impetuous younger brother Henry to become King Henry VIII. This arguably led to the English Reformation.

A relatively benign relationship among rats, fleas, and *Yersinia pestis* becomes the Black Death when humans inadvertently find themselves caught in the cycle.

A mysterious disease, the English sweating sickness, swept England in the 1500s, then disappeared. Some suspect it might have been caused by a hantavirus.

The hantavirus, in contrast, is not spread from person to person. Outbreaks like that seen in the United States have now been recorded elsewhere, but none have been of the same magnitude as the plague during the Middle Ages. An outbreak in Argentina with a closely related hantavirus has been particularly worrisome, however. Delia Enria, an Argentinian physician and virologist, found evidence of the virus being transmitted from person to person, just as the plague bacterium was. For now, the Argentinian strain of hantavirus that eliminated the mouse intermediary has disappeared, and health care professionals have breathed a sigh of relief. Events like this still serve to remind of us of the small differences that distinguish microbes that infect us.

As with most new diseases, the emergence of plague or hantavirus is a complex event. The discovery of the factors involved in the emergence of a new disease often reads like a dramatic detective story.

A Pathogen Changes Tactics

September, 1996, Bariloche, a quiet resort town in southern Argentina. And site of another hantavirus outbreak. But this outbreak had a new twist. One of the physicians treating patients in Bariloche became ill. He traveled 1,000 miles to Buenos Aires to see a specialist. The specialist became ill, and then the specialist's wife became ill. All three of them died, and all three were diagnosed with hantavirus pulmonary syndrome.

But the specialist and his wife in Buenos Aires had no known exposure to the rodents that carried the virus, and they had not traveled to the Bariloche region.

Delia Enria, an Argentinian physician and expert in rodent-borne viral diseases, began to suspect human-to-human transmission. This was a phenomenon that had not been seen before with any of the hantaviruses, and the implications, if true, were frightening. It suggested that a new hantavirus had appeared, one that no longer had to rely on its rodent host for transmission.

Enria contacted her old friend and collaborator at the CDC, C. J. Peters. Although he was skeptical of the story, he realized that if it was true, it was serious. A small CDC team was dispatched to Argentina to help in the investigation. What they found offers us a chilling view of the incredible ingenuity of the microbial world. After ruling out all other means of transmission, Enria and Peters were left with the only possible conclusion. The hantavirus, which they dubbed Andes, had indeed been transmitted from human to human, providing it with a far more efficient means of using us as a major reproductive boost. This meant that avoiding rodents had no meaning, and that the virus could hop on a plane with us and be carried anywhere in the world in a matter of hours.

Delia Enria

This time we were lucky. No more cases were discovered. The virus had burnt itself out quickly and then disappeared. But the event serves as a reminder to us that we are locked in a perpetual dance with the microbial world, one in which we do not always have the lead.

A Modern Plague

Twenty-five years ago another new disease appeared. This disease killed *all* those who got it, but because the first victims were from the ranks of those stigmatized by society, the disease spread and reached epidemic proportions before we reacted. Acquired immunodeficiency syndrome (AIDS) was a disease that challenged our prejudices, but ultimately this crisis revealed the power of both the political and medical institutions in the United States and elsewhere.

Germaine Hanquet

The industrialized nations threw the full weight of their scientific and public health skills into understanding and treating the disease and into controlling the spread of the virus that causes it, the human immunodeficiency virus or HIV. Although there are still almost a million people infected with the virus in the United States, and another 1 to 2 million in Europe, many people think we have contained the epidemic.

But have we? The devastating impact of a new pathogen like HIV emphasizes important differences between the developed and the developing world. At this writing, AIDS has achieved pandemic proportions in sub-Saharan Africa. The devastation that it is causing there has not been witnessed since smallpox was introduced into the Aztec population in the 16th century.

In Botswana, the virus has dropped the average life expectancy from 61 years to 47. In Zimbabwe and other countries, one out of every four to five adults is infected with HIV. Of the 30 million people worldwide currently infected with the virus, 26 million reside in one of the 34 sub-Saharan countries. The deadly disease has changed the face of the continent, much as the Black Death changed Europe in the 14th century.

Africa is not alone. Germaine Hanquet, a Belgian physician with Doctors Without Borders, has experienced the differences that economic development can make on two continents. Hanquet saw the effects of HIV while working in Africa. Now she works in Tegucigalpa, the capital of Honduras, where she is translating her experience in Africa into a project targeted to preventing AIDS among the 5,400 street children who live there.

"The reason I'm now here in Honduras working on an AIDS program is that Honduras is a country with a huge AIDS problem. You have 60 to 70% of the cases of AIDS in Central America occurring in a small country which is only 17% of the region.

"It's urgent to do something. AIDS is affecting the entire country. It's a disease you can prevent, so when you're a doctor, it's a disease you feel you can do something about."

Jumping Species

Scientists now believe that HIV evolved from a similar virus that infects chimpanzees. Named the simian immunodeficiency virus, it does not cause a fatal disease in chimps. We don't know yet whether the virus changed as it changed hosts or whether important differences between the chimps and humans account for its lethal effect on us. Answering that question may give us an important insight into how such an event can occur.

Unwitting Hosts

Many emerging diseases are the result of changes in the way we live. Our expanding human population leads to two of the most important changes—increased urbanization and movement into previously remote areas.

More and more, we live in densely populated urban and suburban worlds. By living closer together, we make it easier for microbes to pass from one of us to the next or from a domesticated animal to us. This is particularly true where urban development has occurred without benefit of adequate sanitary systems and clean drinking water. Contaminated water supplies lead to the spread of microbes that cause diarrheal illnesses, including *Salmonella,* pathogenic *E. coli,* the hepatitis A virus, and a variety of intestinal parasites.

When we seek the beauty and solitude of pristine surroundings for recreation or land development, we are straying into new environments with new microbes. Lyme disease is a well-known case. The mothers of children in Old Lyme, Connecticut, were the first to bring Lyme disease to the attention of the medical community. They noticed that many of their children were suffering from what appeared to be juvenile rheumatoid arthritis. We now know the infectious origins of this cluster of disease, along with an appreciation of how it emerged.

In the 1800s, settlers cleared the old-growth forests of the eastern United States for agricultural development. This did not favor the deer or their natural predators, all of which began a population decline. As agriculture moved west to the Great Plains, new forest growth returned. The new forest was dense with undergrowth that favored the deer but not their predators, and the deer proliferated.

We began to build houses and hike into the newly forested areas with enlarged deer populations, presenting a new host for the microbe that causes Lyme disease, a spirochete called *Borrelia burgdorferi.* The spirochete goes through a complicated life cycle during which it is transmitted between deer and deer mice by a deer tick, none of which are particularly bothered by the microbe. When we unwittingly inserted ourselves in the path of this natural cycle, providing an equally attractive feast for the infected ticks, we opened a new window of opportunity for the spirochete. Lyme disease is now the most common tick-borne infectious disease in the United States.

Borrelia burgdorferi

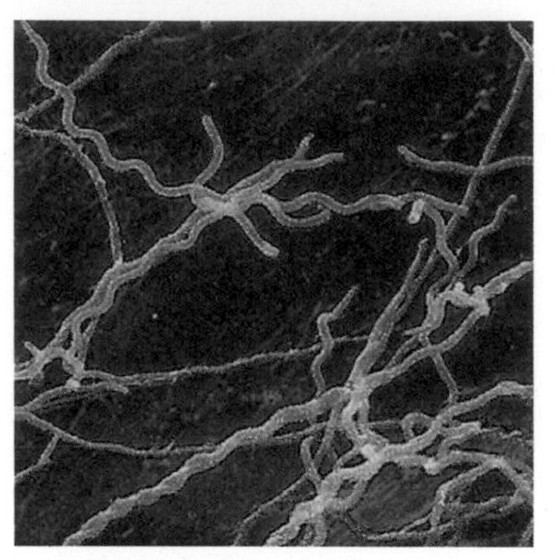

Identity: Spirochete
Residence: Ixodid ticks, white-footed mice, and deer
Favorite pastime: Migrating to the skin to cause eruptions
Activities: A member of The Opportunists, *Borrelia* causes Lyme disease in people who get into its life cycle.

A System Overwhelmed

A sewage treatment plant

The value of clean drinking water and the consequence of living in a large population center were brought home to the industrialized world in a significant manner in Milwaukee, Wisconsin, in 1993. Unusually heavy rainfall overwhelmed the city's sewage system, contaminating the city's water supply with untreated sewage. The contaminated drinking water resulted in 400,000 cases of cryptosporidiosis, a diarrheal illness caused by *Cryptosporidium,* a protozoan parasite.

Resourceful Guests

Microbes evolve much faster than we do. Many reproduce very quickly, giving rise to a new generation as often as every 20 minutes. In addition, they mix and match their genetic information on a regular basis, passing around genes the way we pass around a favorite recipe. Imagine having a baby every 20 minutes while concurrently passing genes for curly hair to all your family and friends. Within hours, people with curly hair would populate the whole neighborhood. In the microbial world, this "great gene shuffle" regularly gives rise to organisms with new options.

New genes can lead to microbes with new options. Some reshuffled genes have information that will give microbes an advantage in invading us. These include devices that help them get the food they need, thwart our body's defenses, and fend off our protective microbial inhabitants. For example, *E. coli* regularly passes the genes for the toxins causing the diarrheal illness we associate with eating undercooked hamburger among its family and friends. And lest vegetarians take too much comfort, scientists have found these gene-carrying toxin producers on everything from apples to lettuce.

If a microbe can break into our body, get food, and reproduce without being detected by our immune system, like a thief making off with the family jewels, it is a happy . . . well, at least a successful microbe.

A proven approach to sneak past our guard is to use a new disguise. The ubiquitous influenza viruses are a prime example. Mutations and gene shuffling occur regularly, changing the viruses' surface enough so that our immune system doesn't recognize them any more. Such changes vastly delay the action of our immune cells, giving the virus plenty of time to reproduce while making us quite sick.

Another proven approach uses a different form of evasion. We have used antibiotics as an effective way of stopping invading microbes since the discovery of penicillin in the first part of the 20th century. Microbes, in turn, have invoked their evolutionary skills to evade or blunt the killing effects of our antibiotics. By developing resistance to many antibiotics and then passing the information around, microbes now regularly escape the controlling effects of our best drugs.

While microbial antibiotic resistance may not give rise to entirely new pathogens, it does empower microbes that we thought were conquered to resurface. From the bacteria that cause tuberculosis and wound infection to the virus that causes AIDS, microbes are rapidly eroding the temporary advantage our antibiotics have given us.

1. Our immune system recognizes, tags, and neutralizes a familiar "flu" virus.

2. Through mutation, the outer coat of the virus changes . . .

3. . . . to one that our immune system doesn't recognize.

An Unpredictable Mother Earth

A powerful factor influencing the emergence of disease is in many respects a wild card. It's the effect that weather extremes (short-term cycles) and climate variation (longer-term cycles) have on changing disease patterns.

Apart from the immediately apparent physical damage to property and loss of lives, weather extremes—hurricanes, tornadoes, and heavy rains that lead to flooding—are opportunities presented to microbes. Such events disrupt critical services, including sewage treatment and water purification. The result is a predictable upswing in water-borne diseases like diarrhea, cholera, and hepatitis.

Climate variation creates weather changes that are maintained over a longer cycle. A familiar form of climate variation is the El Niño cycle, which is responsible for increases in weather extremes such as rain and drought. Scientists are beginning to link such climate variations to shifts in infectious disease patterns. Researchers believe the 1993 outbreak of hantavirus pulmonary syndrome in the United States was linked to an increase in rainfall that resulted in bumper crops of the piñon nut. The piñon is the favored food of the local deer mice. The mouse population increased, and so did the risk of human infection.

Rainy years also favor the proliferation of many varieties of mosquito, an insect that is not only annoying but also an important vector for transmitting disease to humans. Malaria, dengue fever, Rift Valley fever, yellow fever, and encephalitis are all diseases that mosquitoes transmit to people. When the number of mosquitoes goes up, so does the number of infected people.

Understanding how climate and diseases are linked can help us predict and prevent outbreaks. Warm, wet conditions offer fertile breeding grounds for many infectious diseases. Unusually heavy rains overwhelm water systems designed for less extreme conditions, exposing people to diseases like cholera. Heavy rains also lead to more standing water. That plus a warm climate make a more favorable breeding ground for mosquitoes.

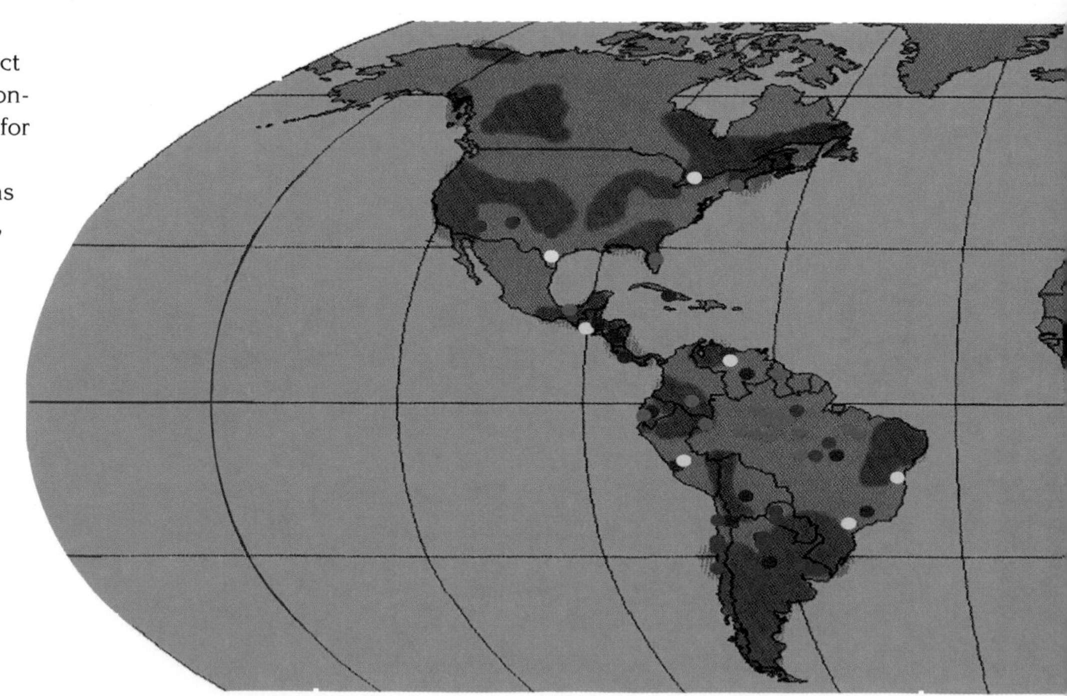

Mosquitoes transmit a variety of diseases to humans, including malaria, dengue fever, yellow fever, and encephalitis. When the climate is wet and warm, the mosquito population explodes, increasing our probability of becoming a mosquito's meal and the recipient of a microbial pathogen.

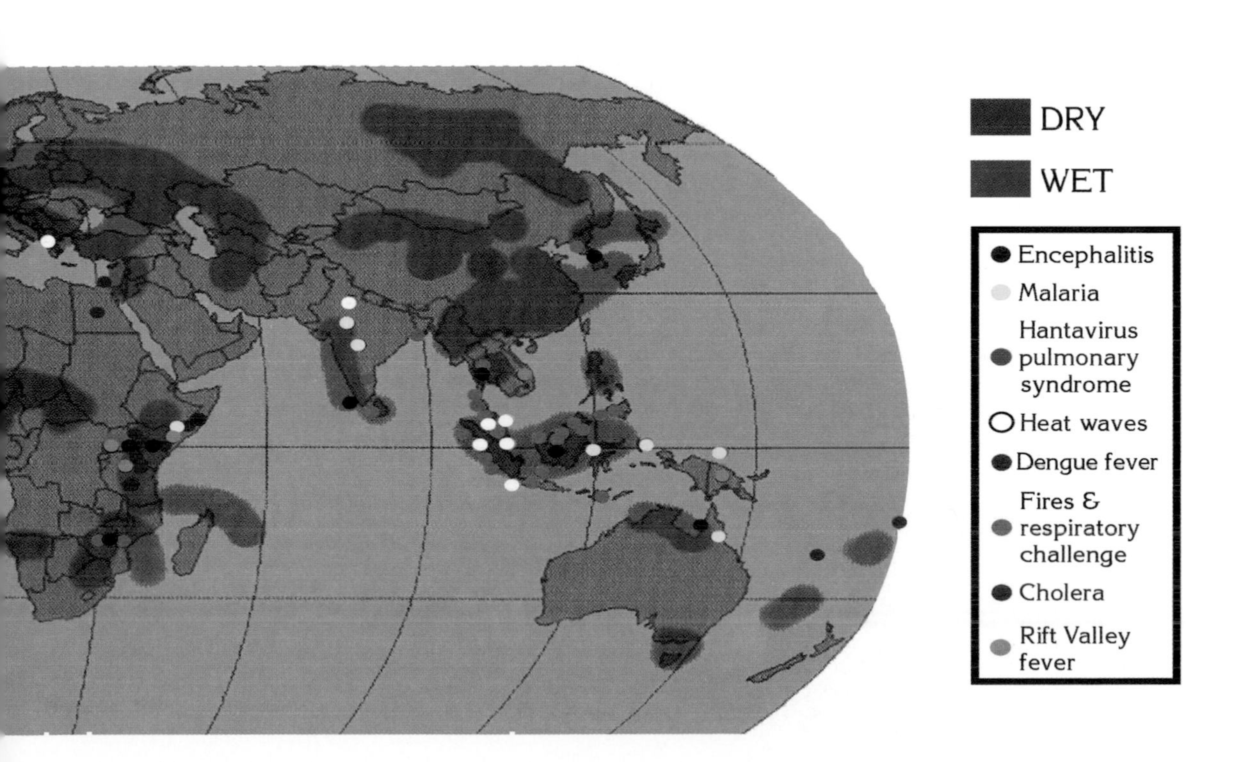

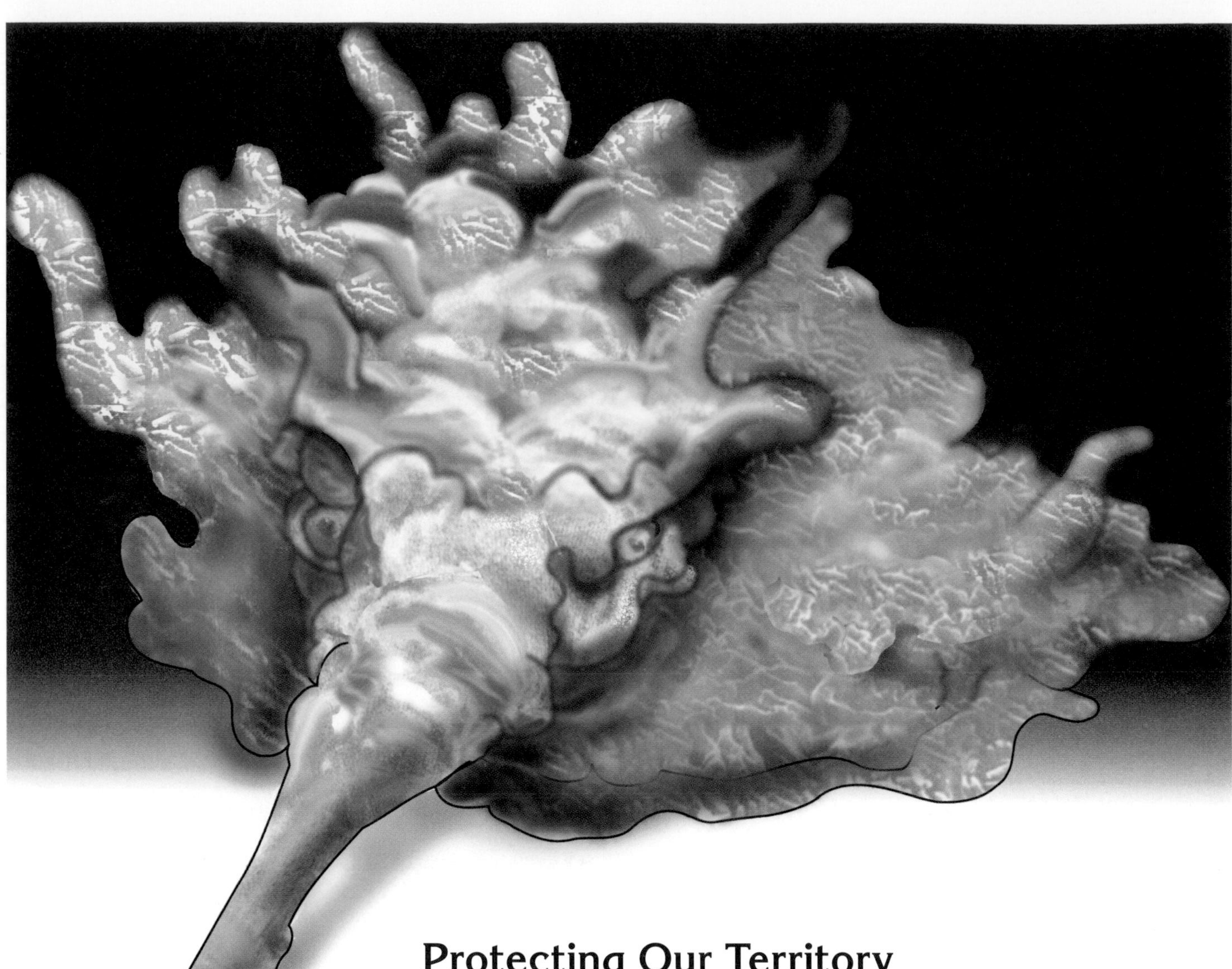

Protecting Our Territory

Since we evolved with and among microbes, it should come as no surprise that we have developed devices for keeping the upper hand. We protect our territory, and any microbe wishing to infect the human body faces a series of formidable challenges.

As we have seen, our first line of defense includes our skin and mucous membranes and the acids, enzymes, and damaging chemicals that bathe them. These defenses present an impressive barrier, but many microbes still find their way through our "Maginot Line." Those that do will encounter our powerful internal defenses—our immune system.

The immune system provides a multileveled defense. The genius of its design lies in its progressive sophistication in recognizing and targeting invaders. At the first level, the system simply identifies and attempts to kill cells that are not part of our body—distinguishing between self and nonself. At higher levels, the system invokes more sophisticated recognition tools to identify a specific invader from among millions of possibilities, and then employs explicit measures to kill the intruder.

How it does this is a remarkable story.

1. Complement proteins (pink) circulating in our bloodstream (among the blood cells) . . .

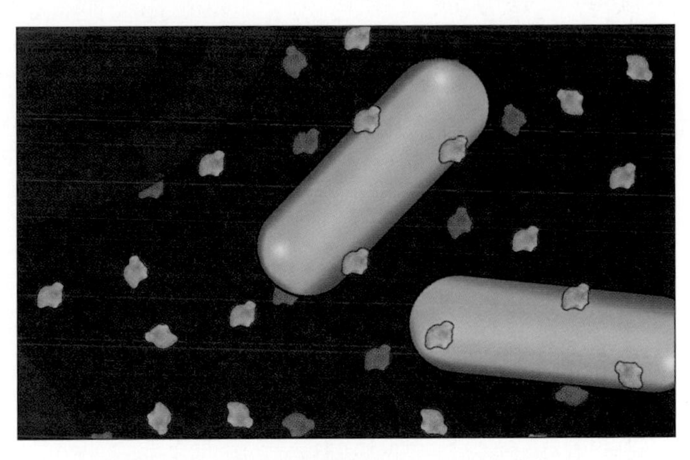

2. . . . latch onto bacteria that have gotten inside us.

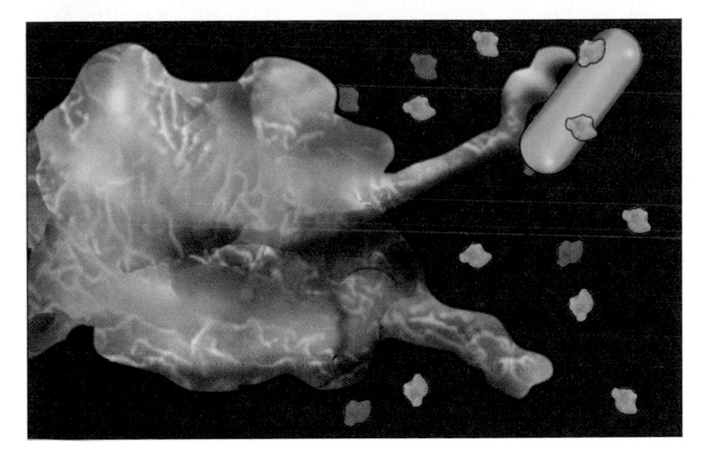

3. The invaders, tagged by the complement proteins on their surface . . .

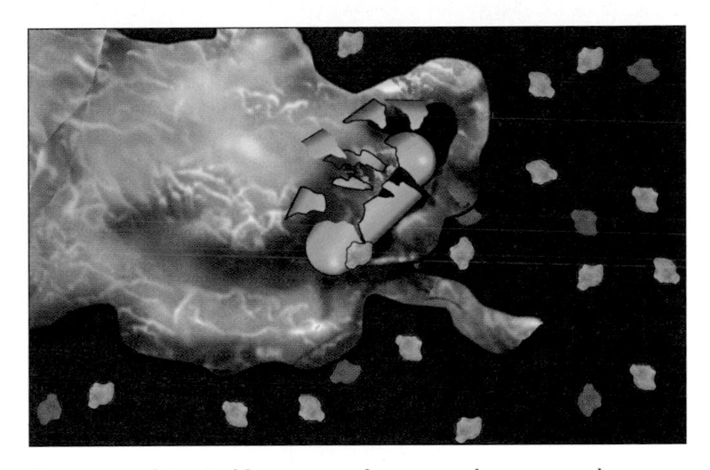

4. . . . are detected by macrophages and consumed.

Skirmishes at the Border

Initial clashes with some microbes—our familiar enemies—are like engagements between two old generals. We have learned each other's strengths and weaknesses from a long history of battles. We also meet microbes that our immune system has not encountered. Unfamiliarity can slow our response, allowing these strangers to slip past our front-line defenses.

Regardless of the intruder, the immune system first relies upon a set of proteins floating continuously through the bloodstream to detect microbes that have crossed our barriers. These proteins attach to the invaders, triggering a set of events called the complement cascade.

Each protein in the cascade performs its own specialized function in the initial battle. Some join together, stick onto the invading microbe, and punch holes through its surface. Others attach and provide a handle for patrolling macrophages—"big eater" cells. These professional eating machines encircle the intruders, gobbling them up.

Many of these encounters are minor skirmishes we never notice. Some are more significant actions that result in a swollen, red, and cell-filled border crossing that we know as inflammation. If the microbes survive, our body turns up the heat, literally. We know this as fever. Fever makes us uncomfortable, but it's worse for some microbes. In higher temperatures, they can't reproduce, which gives us a better chance at preventing further invasion.

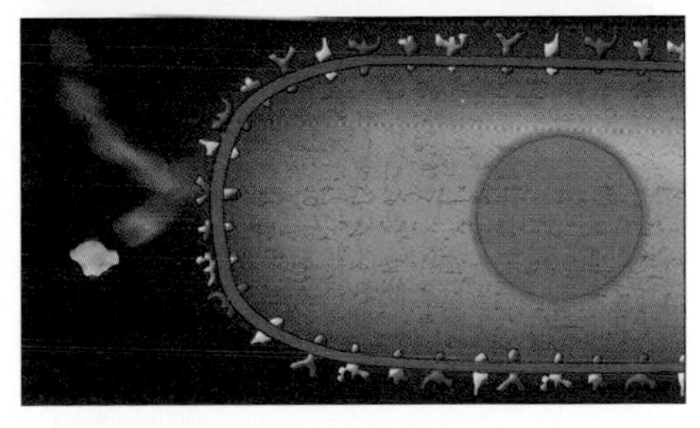

Complement proteins stick to our own cells as well as to foreign cells. But because of differences in the chemical composition of the surfaces, they tend to drop off our cells readily.

The Internal Defense

Pathogens that have eluded the border patrol find themselves in a race to reproduce before the body's next line of defense—the specific immune response—locates, identifies, and kills them. This whole internal defense team is a group of specialists. These cells communicate by chemical signaling so elaborate and tightly orchestrated that some have called this system a "floating intelligence."

Intruder Detected

The first soldiers in the specific immune response team are the roving "big eaters," the macrophage cells that we have already met. Even though their recognition abilities are crude, they usually catch a number of the wily invaders and chew them into pieces. Some of these small pieces end up attached to the outside of the cell, like crumbs on a greedy eater's face. The pieces, called antigens, will serve as critical information for the next group of defending cells. The macrophages say, in effect, "This is what the bad guys look like." The more sophisticated cells of the immune system then waken to the threat and leap into the fray.

Intruder Identified and Removed

The central cells of the specific immune response are the lymphocytes—the T cells and B cells. Unlike macrophages, the lymphocytes do not eat invading organisms. Their role is to recognize and neutralize invading microbes that have evaded the first lines of defense.

In most living cells, one gene makes one protein. But the remarkable lymphocytes, in manufacturing the protein receptors that will appear on their surfaces, randomly cut up certain of their own genes and then rearrange the pieces in a novel way. This means that each new lymphocyte, whether T or B cell, has a unique protein receptor on its surface made from the mix-and-match genes. Each lymphocyte is thus a one-of-a-kind model, its surface receptor different from that of any other lymphocyte.

The lymphocytes, born in the bone marrow at the rate of tens of thousands per minute, join with billions of their kin that circulate in the blood and lymph. Such sheer numbers make it highly probable that at least one of these will find and "fit" (i.e., form a lock-and-key receptor coupling) with the "piece of the bad guy" presented by the macrophage.

T cells orchestrate the internal defenses. Using chemical signals, they help or suppress all the other immune cells' activities. When a T cell binds with an antigen from a pathogen, the T cell starts dividing. One cell becomes two, two become four, then eight, and in a surprisingly short time there are millions of lymphocytes that can recognize the pathogen.

The Attackers

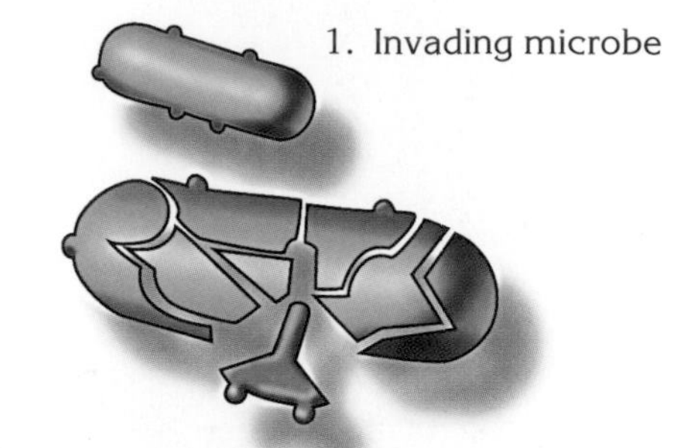

1. Invading microbe

2. Antigen (a piece of the microbe)

The Defenders

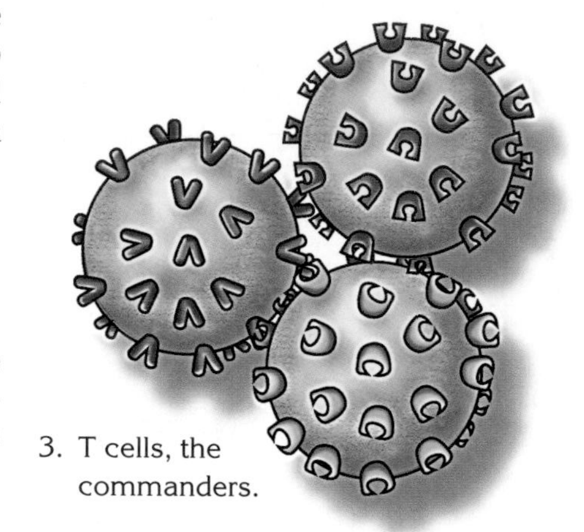

3. T cells, the commanders.

4. B cells, the artillery.

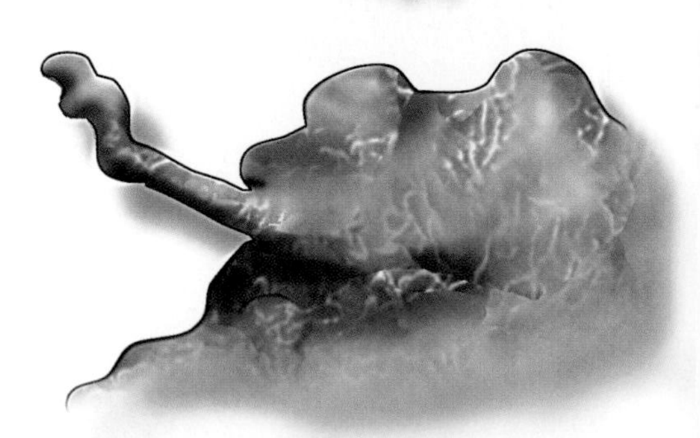

5. Macrophage, the terminator.

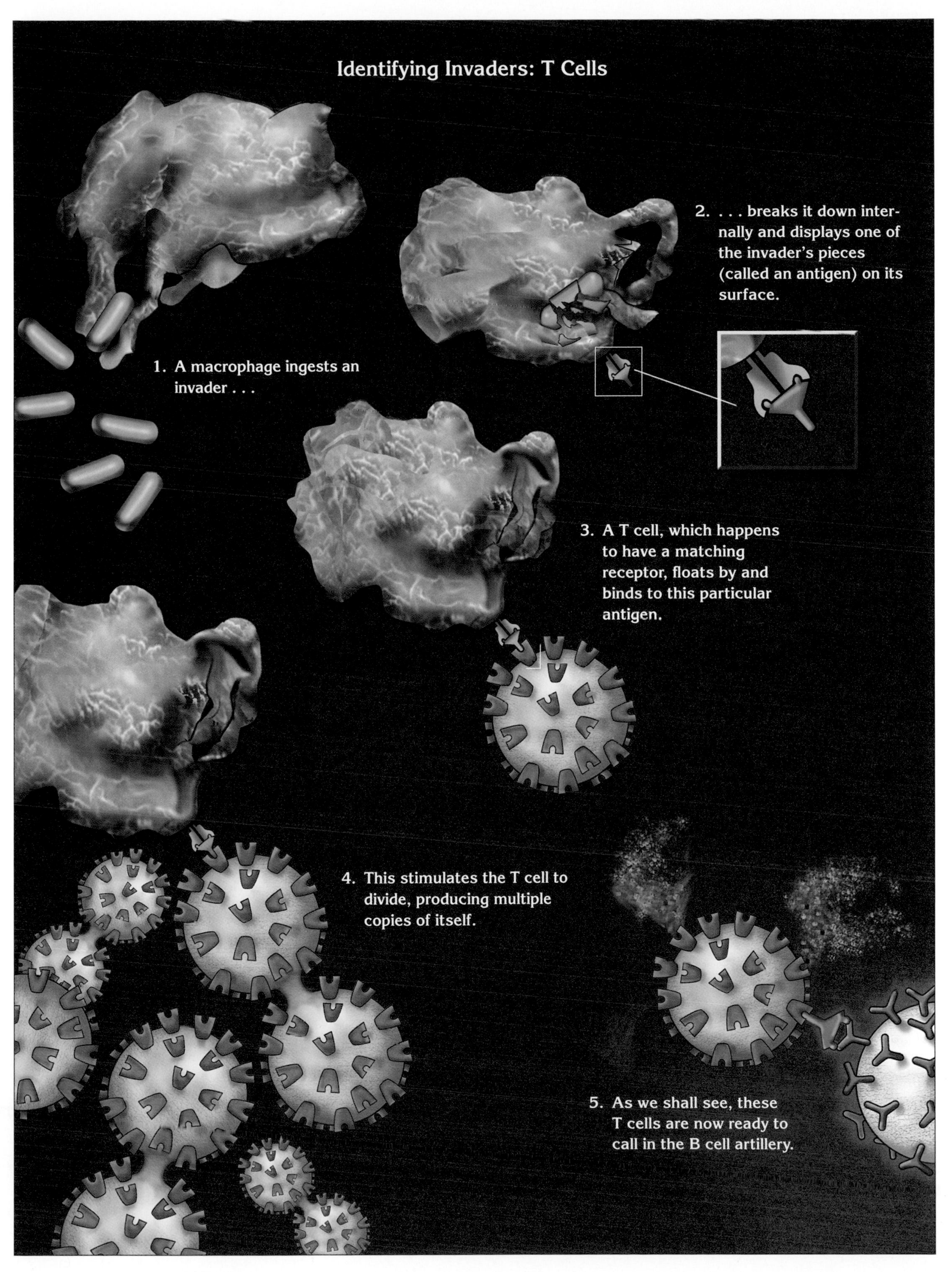
Identifying Invaders: T Cells
1. A macrophage ingests an invader . . .
2. . . . breaks it down internally and displays one of the invader's pieces (called an antigen) on its surface.
3. A T cell, which happens to have a matching receptor, floats by and binds to this particular antigen.
4. This stimulates the T cell to divide, producing multiple copies of itself.
5. As we shall see, these T cells are now ready to call in the B cell artillery.

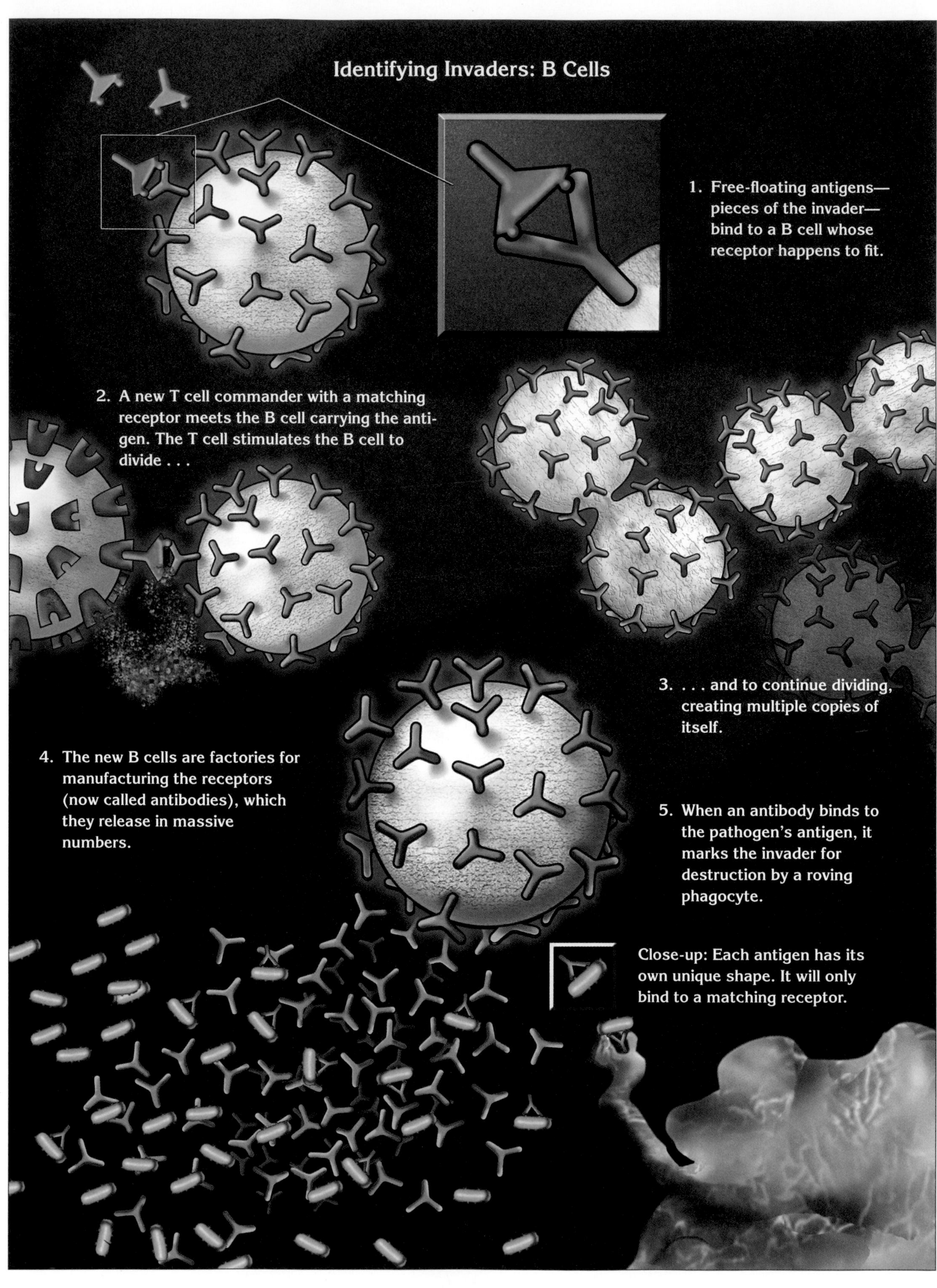
Identifying Invaders: B Cells
1. Free-floating antigens—pieces of the invader—bind to a B cell whose receptor happens to fit.
2. A new T cell commander with a matching receptor meets the B cell carrying the antigen. The T cell stimulates the B cell to divide . . .
3. . . . and to continue dividing, creating multiple copies of itself.
4. The new B cells are factories for manufacturing the receptors (now called antibodies), which they release in massive numbers.
5. When an antibody binds to the pathogen's antigen, it marks the invader for destruction by a roving phagocyte.
Close-up: Each antigen has its own unique shape. It will only bind to a matching receptor.

B cells perform a different function. B cells release their receptors, called antibodies, to float freely in large numbers. When an antibody locks onto an antigen on the microbe's surface, it makes the invader fully visible for the next available microbe-eating phagocyte that charges into the fray.

Intruder Remembered

When all works as it should, our specific immune response wards off the microbial incursion, and we live to see another day. And, astonishingly, certain cells within the specific immune response team retain the memory of the intruder. The memory is called immunity, and once established, our specific immune system can gear up much more rapidly the next time the same microbe slips by our first line of defense.

We take advantage of our immune system's ability to develop memory when we undergo vaccination. Vaccines contain microbes or parts of microbes that we have "defanged." Measles vaccine contains measles virus and tetanus vaccine contains tetanus toxin, but we have modified each so it doesn't cause disease. The cells of our immune system, however, can't distinguish between vaccine and the real thing. By confronting the system with an ersatz pathogen, we gain the memory without suffering through the infection. When the real thing does come along, our specific immune defense system is ready and prepared to swing into action before the invader ever has a chance.

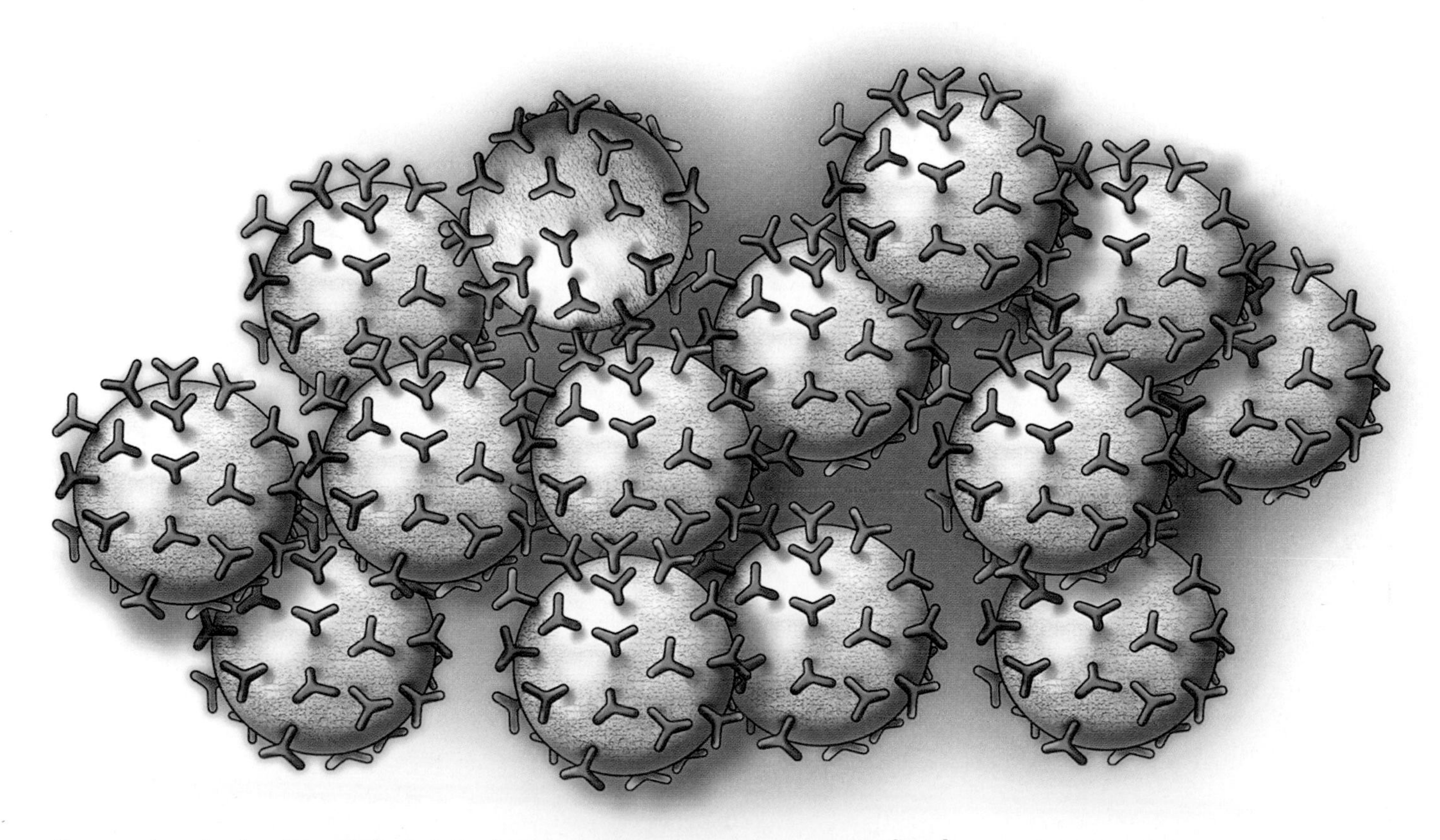

Once an invader has been defeated, the immune system has an insurance policy. It leaves a population of long-living B cells and T cells that match the antigens from the invader. They are ready to respond quickly should the invader ever reappear.

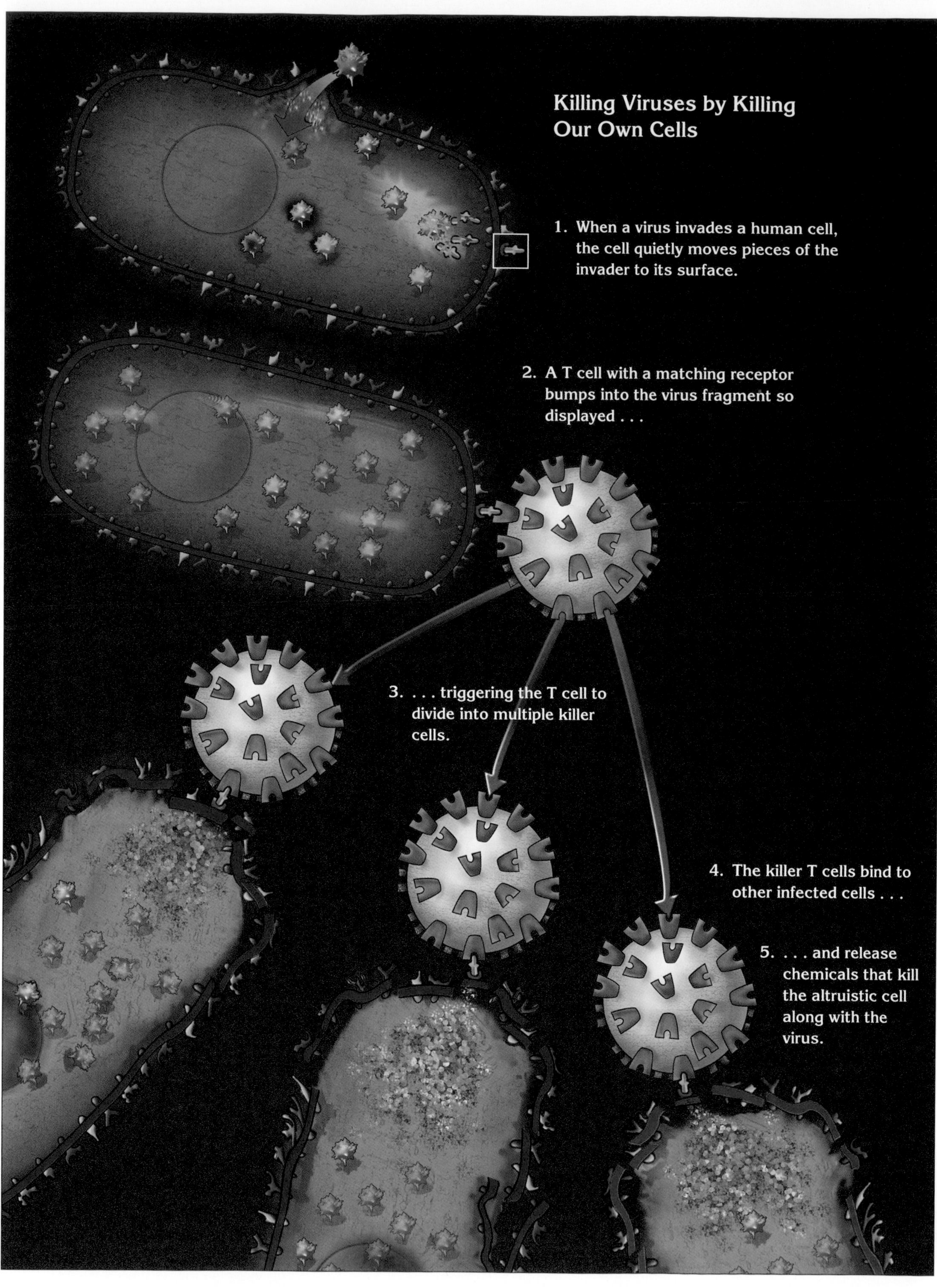
Killing Viruses by Killing Our Own Cells
1. When a virus invades a human cell, the cell quietly moves pieces of the invader to its surface.
2. A T cell with a matching receptor bumps into the virus fragment so displayed . . .
3. . . . triggering the T cell to divide into multiple killer cells.
4. The killer T cells bind to other infected cells . . .
5. . . . and release chemicals that kill the altruistic cell along with the virus.

Outwitting the Immune System

Obviously, there are times when this complicated defense system is outwitted or just can't recognize the invading microbe. Fortunately for us, the system is not outsmarted easily and almost always can recognize an intruder. Otherwise, we most certainly would not have survived as a species.

Evolution is such that a clever immune system will spawn clever microbes. In this evolutionary arms race, different microbes have perfected various successful strategies to get around our defenses.

Some microbes can change the specific bits and pieces on their surface—their antigens—so that the memory cells within our specific immune response team don't remember them from previous encounters. The influenza virus, for example, can change its surface proteins through mutation, which is why it can evade our immune system even after we have been vaccinated. This year's virus may wear a new coat, so it doesn't look like last year's version.

Some microbes cloak themselves in proteins or carbohydrates that look like ours, tricking the immune system into considering them one of our cells. The bacterium that causes strep throat, for example, coats itself in a substance called hyaluronic acid. This same substance serves as the glue that holds human cells together. This clever charade allows the microbe to grow and divide unnoticed for long enough to establish a firm foothold in our tissue.

Other microbes produce a host of substances that kill the macrophages or communicate misleading information. *S. typhi* shuts off the "big eater's" killing mechanisms and uses the cell like a recreational van to travel around our bloodstream. HIV infects certain T cells and wrecks their ability to orchestrate the immune system.

Amid all of these microbes with their devices, the viruses deserve special mention. In order for a virus to replicate, it must first invade a host cell. Viruses are so simple that they lack the ability to reproduce on their own, so they commandeer the machinery of the cells they infect in order to multiply. This takeover puts them out of reach of our immune system. Some of these invasions result in a cellular version of altruism, as the invaded cells display viral proteins (antigens) on their own surface. The viral proteins alert the cells of the immune system that a foreigner is lurking within, but the end result is often fatal for the infected cell, as T cells surround and kill it.

Certain viruses, like the herpesviruses, even go so far as to integrate their genetic material with our own, becoming so "quiet" that they arouse no suspicions. Once established, the virus can and usually does remain with us for life. Some three hundred remnants of viral genes can be identified in our own genetic material, their genes a reminder of long-ago invasions.

Polyomavirus

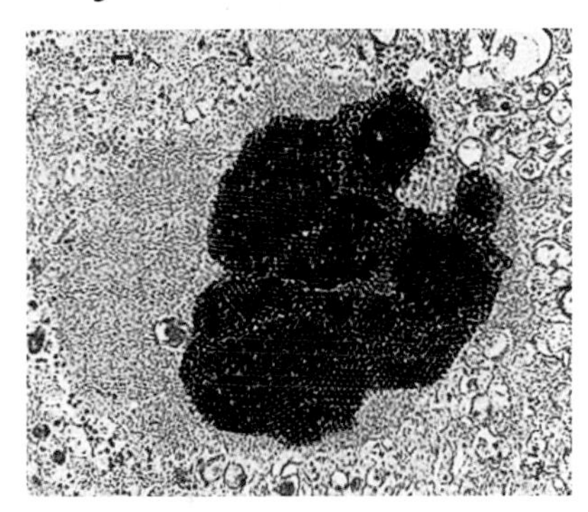

Identity: Virus
Residence: Inside the kidney cells of humans
Favorite pastime: Sitting quietly inside its cell
Activities: A member of The Opportunists, the polyomaviruses prevent the cells they infect from displaying a piece of the virus on their surface, which allows the virus to go undetected by killer T cells; when the host's immune system isn't working properly, the virus can break out and cause disease.

Damage by Friendly Fire

We cannot blame all the damage during an infection on the invading microbe. Our own immune system almost always causes collateral damage in its attempts to thwart an invader. In fact, in some cases, our own immune cells and their products create almost all the damage. *Streptococcus pneumoniae,* a bacterium that causes pneumonia, does little but evade being eaten by phagocytic cells. It is the fury of the phagocytic cells' frustrated attack, releasing chemicals and calling in reinforcements, that ultimately causes the majority of the damage to the lung.

Hantavirus pulmonary syndrome is an extreme example of the fury of the immune system. The extensive damage and fluid accumulation that occur in the lungs of its victims are really a consequence of an overzealous immune attack against the virus. It's this overzealous response that is a major factor in the lethal effects of the disease.

Victims of hantavirus pulmonary syndrome become infected after they inhale dust contaminated with the dried urine and feces of infected deer mice. The virus enters the bloodstream from the little air sacs in the lungs called alveoli. From there, the virus attaches to and invades its target, the cells that make up the lining of the tiny blood vessels that surround the alveoli. These little vessels are responsible for exchanging the carbon dioxide from our own cells' respiration for the oxygen in the air sacs. Once inside the target cells of the vessels, the hantavirus takes over their machinery and makes multiple virus copies.

The infected cells, however, send out a signal. They display some of the virus's proteins on their cell surface. The macrophages and T and B cells tumbling by recognize the signal as trouble, and the immune defense system goes into high gear. The immune cells secrete potent chemicals that inactivate the virus and call in reinforcements. As the inflammatory battle escalates, the chemical barrage weakens the blood vessels, allowing plasma, the fluid part of our blood, to leak out. Before long, the plasma has filled the alveoli of the victim's lungs, leaving no room for air.

This may seem paradoxical, since our host defenses are clearly designed to protect us. However, inflammation creates its own aftermath. Our body is the battleground and inevitably suffers the consequences of the action. While the immune system is valiantly fighting off some intruders like the hantavirus, it may also be blowing holes in our own cells and generally creating an ugly mess as a result of "friendly fire."

Hantavirus Pulmonary Syndrome: Friendly Fire

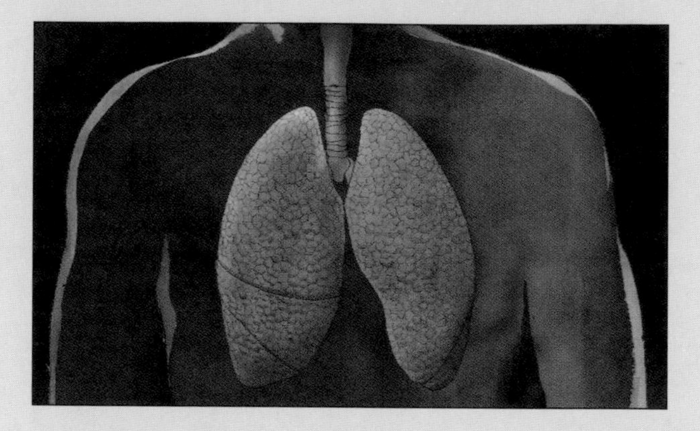

1 Victims inhale the virus . . .

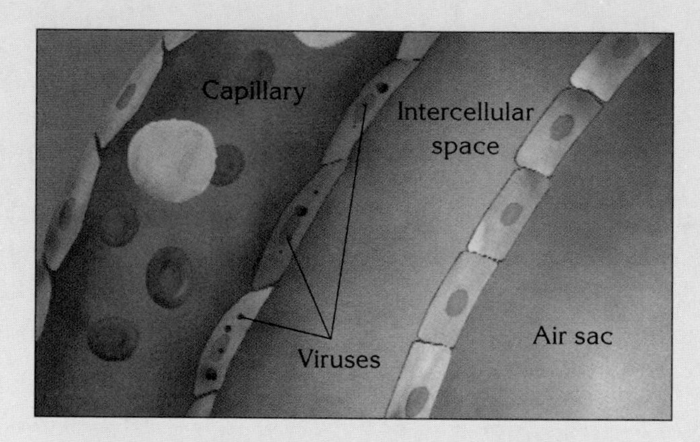

4 . . . and takes over the cell's machinery in order to replicate. The infected cells display the virus's proteins on the surface, signaling trouble.

7 The leaking plasma fills the lung's tiny air sacs . . .

2 . . . which enters their bloodstream.

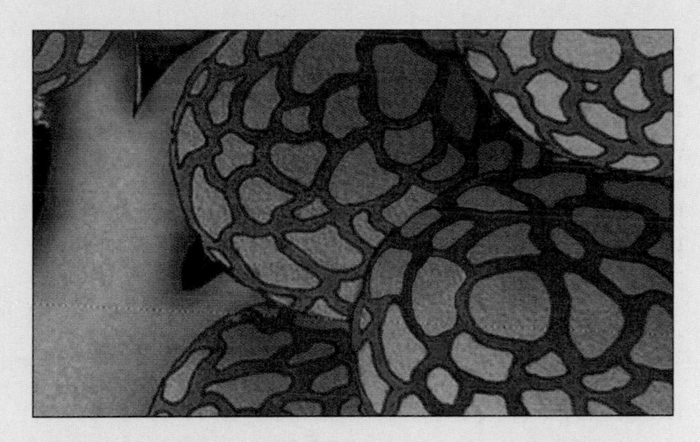

3 From there, the virus invades the cells of the capillaries that surround the lung's tiny air sacs . . .

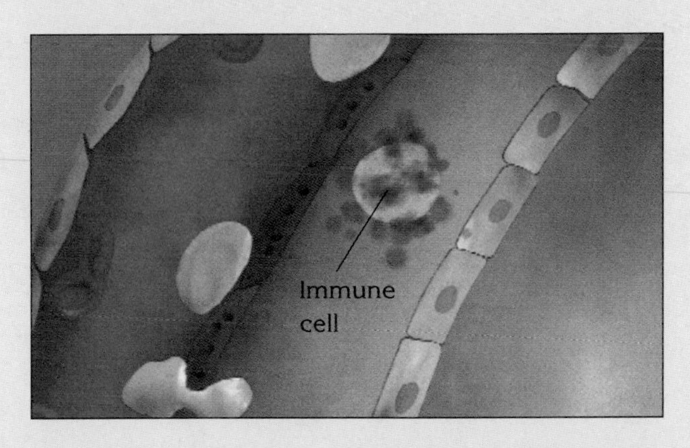

5 Cells of the immune system go into high gear, secreting potent chemicals that inactivate the virus and call for reinforcements.

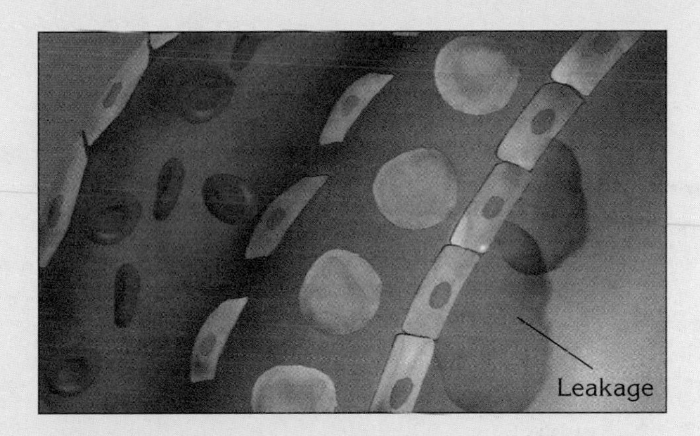

6 The ensuing chemical barrage weakens the vessels, allowing plasma, the fluid part of the blood, to leak out.

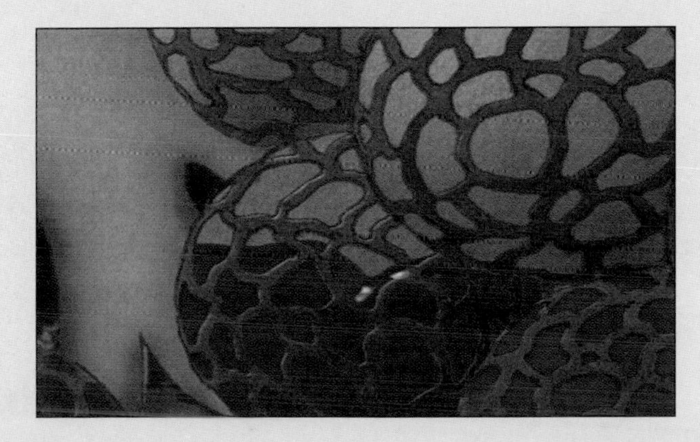

8 . . . leaving no room for air . . .

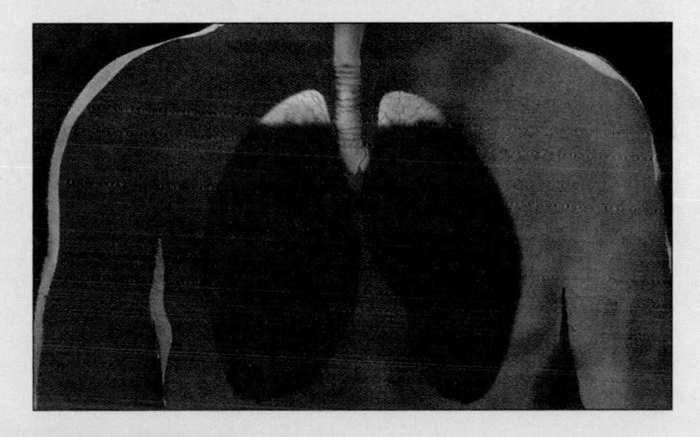

9 . . . ultimately destroying the lung's capacity to function.

Our Battle To Control Infectious Diseases

Until the 20th century, infectious diseases were the leading cause of death in both industrialized and developing countries. The general life expectancy was around 45 years. Only 10% of the population lived to the age of 60.

Today in the industrialized world, the average life expectancy has risen to over 75 years. Pneumonia and influenza, two common infectious diseases, have moved from first to fifth place on the list of leading causes of death, falling far behind heart disease, stroke, cancer, and accidents. Tuberculosis and gastroenteritis, once second and third among causes of death, no longer appear in the top ten. This remarkable shift has come about primarily because we have learned much about how to prevent and treat infectious diseases.

Shigella

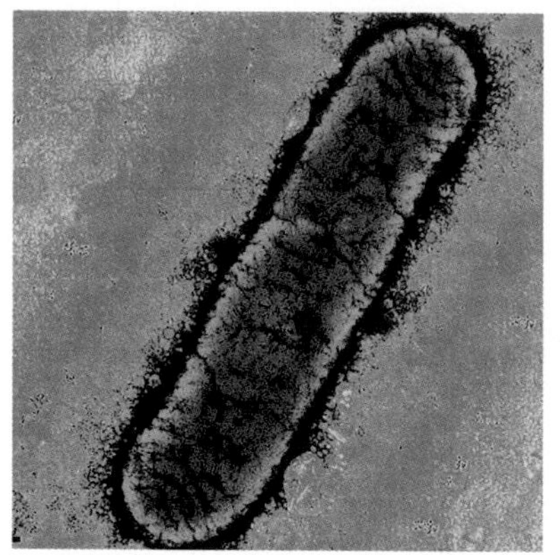

Identity: Bacterium
Residence: The small intestine of humans
Favorite pastime: Contaminating food and water
Activities: A member of The Human Pathogens, *Shigella* causes dysentery, a most unpleasant form of diarrhea.

Clean Water

Once we learned in the 1800s that germs cause disease, we began to identify specific microbial pathogens and the routes they took to bring about human disease. In the United States then, as in many developing countries today, the water supply represented a major route for the spread of germs. Gastroenteritis—and the resultant diarrhea—was the third largest cause of death and a leading cause of disease. *Salmonella, Shigella,* and certain strains of *E. coli*—microbes that cause diarrhea—reach us through drinking water and food that have been contaminated by human and animal fecal waste. We were able to reduce the disease caused by these microbes by cleaning up our water supply and learning to wash our hands. International agencies like the World Bank and the United Nations are trying to achieve the same results in developing countries around the world, where clean water is in short supply.

Disease Detectives

With our advancing understanding of disease transmission and its prevention came the need to track microbial pathogens ever more carefully. Many scientists and physicians today work on the front line in the investigation and control of disease outbreaks. Many of them are members of an international network of government agencies and laboratories. The U.S. Public Health Service, with agencies including the Centers for Disease Control and Prevention (CDC), the National Institutes of Health, and the Food and Drug Administration, along with the network of state laboratories and health agencies, coordinates our efforts in the United States. These agencies are responsible for maintaining surveillance and enforcing the standards that keep us free from the constant epidemics of the past. Other countries have their own counterparts, and agencies like the World Health Organization provide global coordination and policy development.

It was through the efforts of disease hunters and organizations like these that smallpox was finally eradicated from the globe and other diseases, like polio and measles, were substantially brought under control. Such organizations are often the first to respond to the outbreaks of mysterious new illnesses like those caused by Ebola virus, hantavirus, and *Legionella pneumophila.*

John Snow: The First Disease Detective

John Snow is viewed by many as the father of epidemiology, the approach used by modern public health services to track disease. Snow, the son of a farmer, was apprenticed to a Newcastle surgeon when he was 14. In 1831, as he was nearing the end of his apprenticeship, he was sent to help the victims of a cholera outbreak. He became a lifelong student of the disease as a result.

Scientists were convinced at the time that cholera was caused by "miasmas and mephitic vapors." Snow became convinced that the disease was transmitted by something in the water supply. In August 1854, Snow was in London when a large outbreak of cholera occurred in the Golden Square area, a part of the city housing some of London's poorest citizens. By the end of September, 616 persons had died of the disease.

Snow was determined to track down the cause. He obtained a list of the victims and where they had lived from the General Register's office. Using a large map, he marked the residence of every victim who had died during the week ending September 2. The marks on the map revealed that most of the deaths had occurred within a circle with a 250-yard radius, at the center of which was the Broad Street water pump.

At least one victim lived outside the circle. Snow tracked her down and learned that she used to live near the Broad Street pump. She liked the taste of the water so well that she often returned to her old neighborhood just for the water. Snow had to conclude that water was the heart of the problem.

Ultimately, he got permission to remove the handle of the pump so that no one could get drinking water from it, and the outbreak disappeared. Shortly thereafter, workers excavated the pump and found the water to be contaminated with sewage from the adjacent houses.

Snow's methods beautifully illuminate the way science goes about solving problems. His methods furthered the germ theory of disease and established him as something of a hero among modern epidemiologists, who still use his methods of tracking disease today. And if you visit London, you can still see the location of the infamous Broad Street pump. It's marked by red paint on the curb near the John Snow Pub at the corner of Broadwick and Lexington Streets, now in the heart of one of London's more fashionable shopping areas.

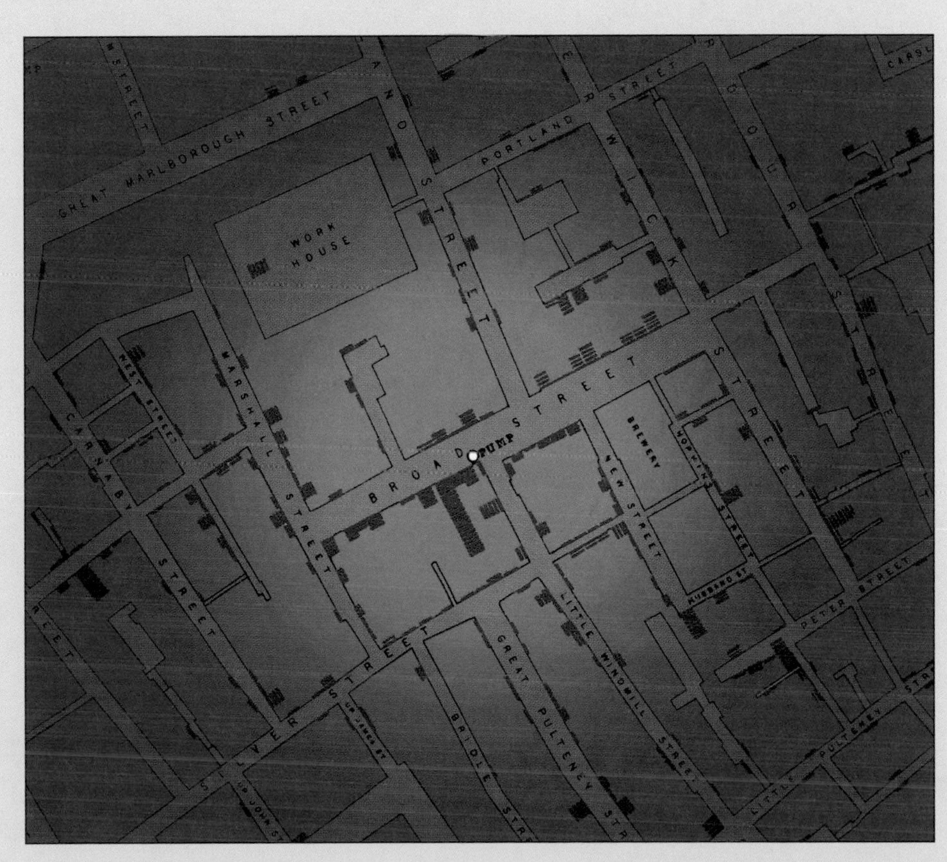

John Snow mapped a radiating circle of cholera victims and found the now infamous Broad Street pump at the center.

Native Wisdom

As soon as the outbreak of hantavirus pulmonary syndrome (HPS) in the Four Corners region of the southwestern United States was recognized, the full force of the Public Health Service, including the CDC, turned to discovering its source. An important clue came from traditional healers—Navajo medicine men and women. Dr. Ben Muneta was a physician and member of the epidemiology team investigating the outbreak. He is also the son of a Navajo medicine man. When members of the tribe were being interviewed, Muneta sensed that they were holding back information. He met privately with the local healers and, to his dismay, discovered that the disease they were tracking had a centuries-long history among the Navajo.

Ben Muneta

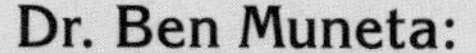

Dr. Ben Muneta:

"What they told me was quite astounding. There was knowledge of an illness that was carried by the deer mouse that is so ancient it's embedded in the Navajo creation story." The Navajo believe that the deer mouse carried the plant seeds to their tribal lands to create the ecosystem—the piñon juniper—that now exists there. Because of their contribution, the deer mice are revered and respected. Even today they are considered important to maintaining the piñon juniper ecosystem, and consequently it is a very bad thing to destroy a deer mouse. On the other hand, there is fear about this rodent.

"It was explained to me by the healers in this manner. Humans occupy a part of this land; they have the day cycle when they are active. The deer mouse is active in the night cycle, so we are two species that never interact although we occupy the same land. At night we seal our homes to keep the deer mouse out, because if the deer mouse comes into the house and sees food scattered on the floor, he becomes angry. The seeds are the essence of creation and you are destroying a life. [The healers] say that they will take the best in that household to die for this transgression.

"The other thing the traditional healers told me was that the way the deer mouse will take a human life is that they have this illness in their urine and droppings that they will pass on to humans. The way it passes is through your eyes, or through your mouth, or through your nose. They described the cycles of this illness. It had to be two good years of heavy snowfalls and rain where the grasses would grow, and then the deer mice would proliferate and then the outbreak would occur during the spring planting season. The medicine people knew where the illness was coming from, they knew the mode of transmission, and they also had recommendations for prevention."

Responding to the convergence of information, the epidemiology unit began trapping small mammals and rodents around the houses of the HPS victims. They found the hantavirus in the deer mice, just as predicted by the Navajo healers. It was a new strain of hantavirus, and it was identical to the virus isolated from the lungs of infected patients. The year 1993 was the second year of El Niño, the warm current that circulates in the southern Pacific. It affects the weather patterns in the southwestern part of the United States, causing increased snow and rainfall. With the increased moisture, the piñon juniper flourished, producing a bumper crop of piñon nuts, the favored food of the deer mouse. With the abundance of food, the deer mouse population increased. And with the increased mouse population came increased opportunity for people living in the area to come in contact with the mouse and its droppings—and thus the deadly virus.

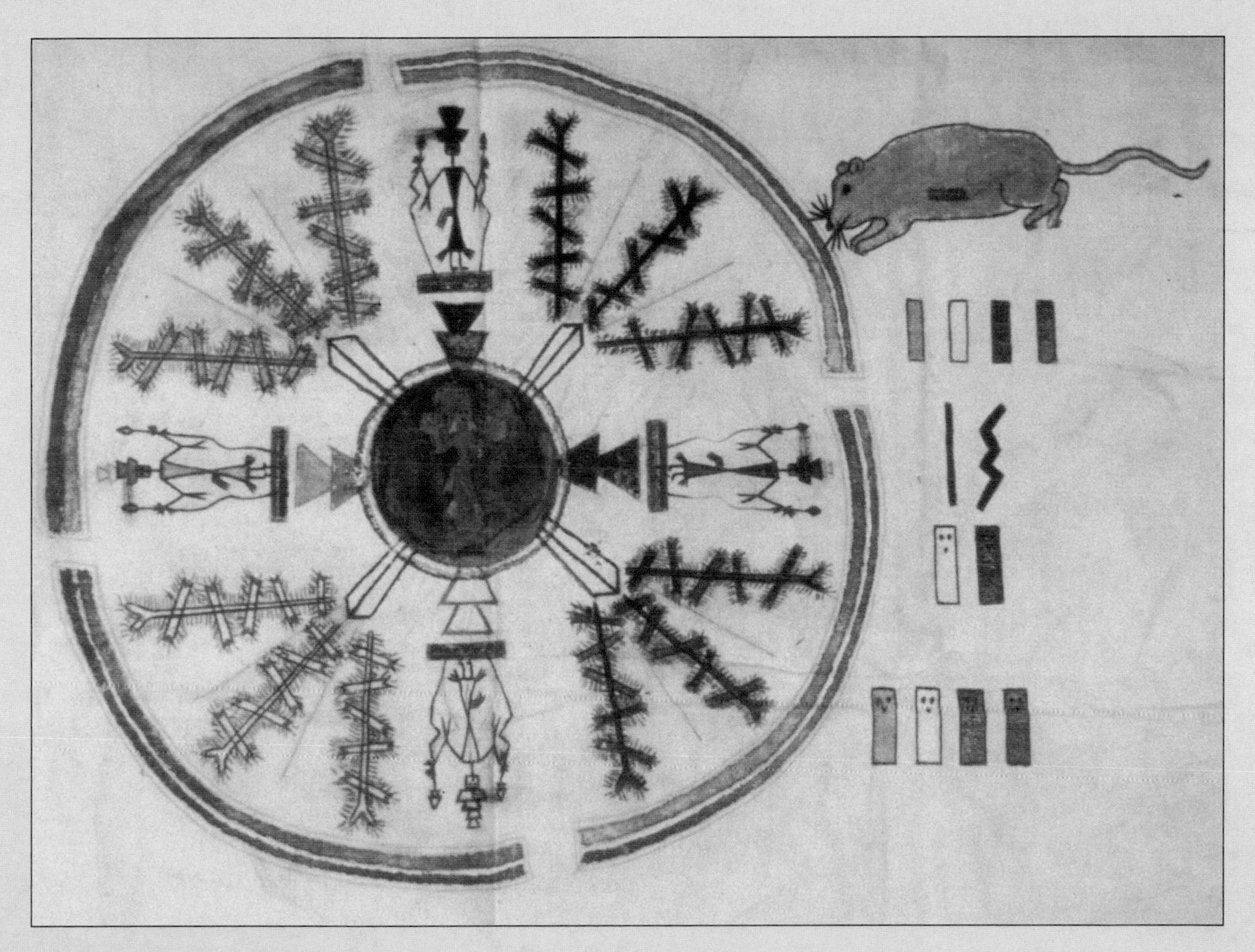

Navajo drawing tells the story of the deer mouse and the hantavirus.

Vaccines

As we identified microbes and associated them with specific diseases, we gained another important tool in our drive to prevent infection—vaccines. Vaccines allow us to educate our immune system to recognize microbes by showing it the bits and pieces of the microbes before they invade. Vaccines change an alien microbe into a familiar one, improving the response time in mounting a successful defense.

Immunization—the administration of a vaccine—is a highly effective way of preventing some diseases, like polio, measles, and diphtheria.

The role of immunization cannot be underestimated in its impact on diseases worldwide. We have virtually eliminated measles, mumps, diphtheria, whooping cough, and most recently *Haemophilus influenzae* ear infections and meningitis. We have totally eradicated smallpox and are close to eradicating polio through worldwide vaccination.

In some surprising ways, we are in danger of becoming the victims of our own success. As our collective memory of infectious diseases like whooping cough and polio fades, the rare complication from vaccination looms large. Because of concerns about such complications, some parents are choosing not to have their children appropriately immunized. This poses a significant threat to the public health, since the microbes that cause the diseases are still very much with us. With the appearance of a large number of susceptible people again, we can expect to see the return of diseases we thought conquered.

Recent outbreaks of whooping cough in western Massachusetts and California and measles in Pennsylvania and Texas among children who had not been immunized serve as stark reminders that the risk of the disease is far greater than the risk of vaccination. In the measles outbreak, at least 21 children died.

Vaccines and our immune system

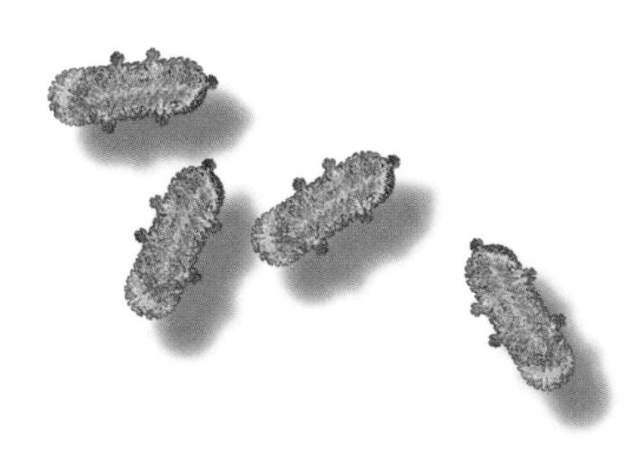

Vaccines may be living pathogens (bacteria or viruses) that have been rendered incapable of causing disease; they may be dead pathogens, or parts of dead pathogens.

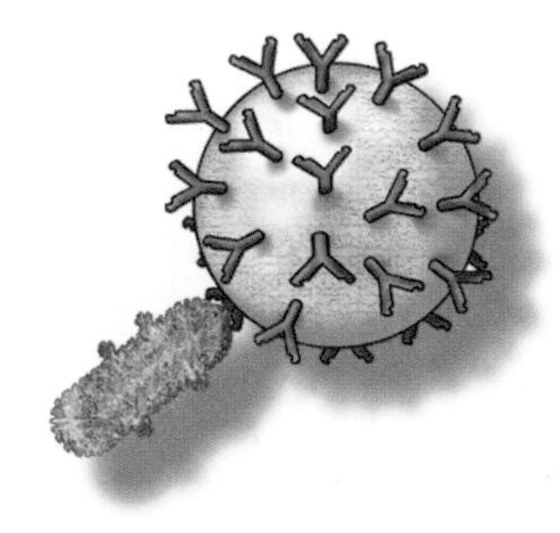

All have the common property of being antigenic, i.e., they bind to their matching receptors on T cells and B cells . . .

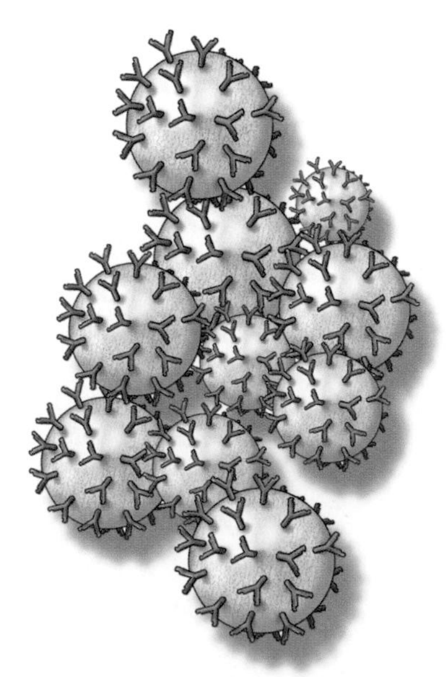

. . . which then multiply into a population of cells ready to respond should the real thing invade.

Antibiotics

Another major contributing factor in controlling infectious diseases was the discovery of antibiotics. Antibiotics are compounds that kill or wound an invading microbe. Before their discovery, many infectious diseases were untreatable.

The first mass production of a useful antibiotic made headline news.

Between the first and second world wars, Alexander Fleming, a Scottish scientist, discovered that the common bread mold *Penicillium* produced a substance that killed certain of the most common bacteria causing wound infection. He named the substance penicillin. Ernst Chain and Howard Florey, a biochemist and a physician working together in Great Britain, recognized the importance of Fleming's discovery, and under the pressure of the impending Second World War, they proved the drug's therapeutic properties in what today would be considered record time. Penicillin was rushed into commercial production, the pharmaceutical industry was transformed, and the age of antibiotics had begun.

With penicillin and sulfa (another antibiotic) available during World War II, the death rate of American soldiers from infectious diseases dropped from 14 out of every 1,000 to less than one per 1,000. This dramatic change was a harbinger of what the discovery would contribute to the overall health and well-being of humanity.

Penicillin and other antibiotics have become the staple of the medical community. They have made many previously incurable diseases treatable, including lobar pneumonia (a major cause of death worldwide), tuberculosis, malaria, and cholera. Antibiotics have given us a tremendous advantage in the give-and-take between our dangerous friends, our familiar enemies, and us.

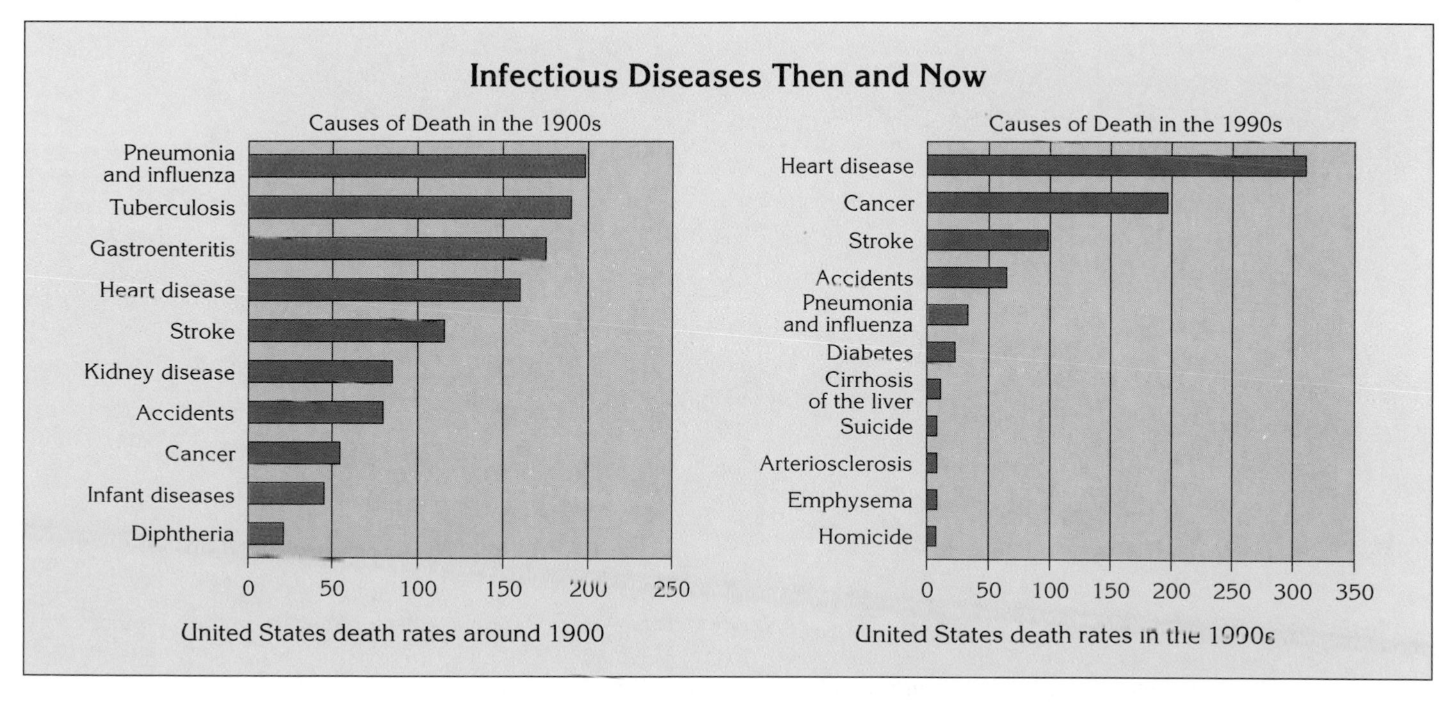

After a century of improved sanitation and antibiotics, infectious diseases in the United States are no longer the number one cause of death.

Our Evolutionary Dance with Microbes

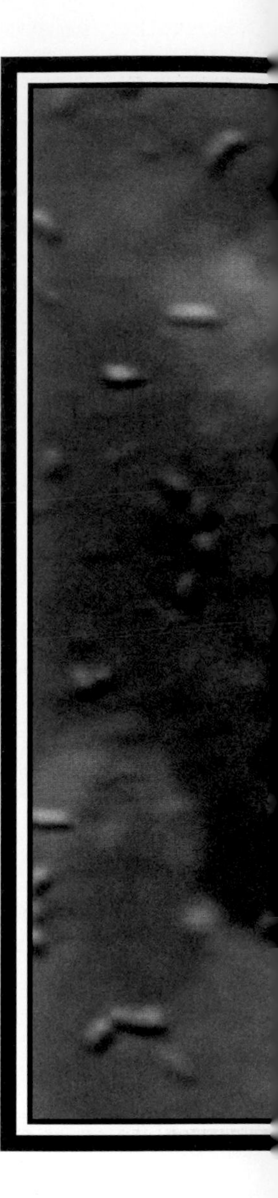

Even in the face of all the challenges, our success in designing sound prevention and treatment strategies might tempt us into thinking that we will eventually eradicate infection. Such a presumption would be a mistake.

The impact of infectious diseases is still staggering in regions of the world where clean drinking water, uncontaminated food, good nutrition, and health care are not accessible. Globally, infection still ranks among the top five causes of death. Malaria kills 1.5 million people a year. Tuberculosis has seen resurgence, claiming over 3 million lives annually. Diarrheal diseases are still the leading cause of infant death worldwide, claiming the lives of between 4 and 6 million children annually.

As a consequence of the genetic ingenuity of microbes, and our own bad practices, antibiotic resistance is now a common problem. A few microbes, like certain strains of *Staphylococcus aureus* and *Enterococcus,* have become virtually untreatable, threatening to return us to an era like the one that existed before the discovery of antibiotics.

This all serves to remind us that we are in an endless dance with microbes. We are intertwined and we will move and change in concert with each other as long as we humans survive. Microbes are our most intimate and oldest partners.

Microbes respond to our every move—to pressures we apply in our desire to prevail. They evolve resistance to our antibiotics and new devices for evading our vaccines. They switch partners when we eliminate a favorite host or present them with a tempting new one.

We respond to their lead—to pressures they apply—with ruthless moves. We design new antibiotics when they develop resistance and new vaccines when they escape our old ones. We eliminate the vectors that carry them from one partner to another.

In the end, our efforts to dominate will almost certainly fail. They have been practicing their moves for close to 4 billion years of evolutionary history. Perhaps new approaches that rely on understanding and entering into their microbial conversations can replace our present strategy to eliminate them. Anticipating and turning aside the most sinister elements of our ongoing tango just might allow us to move forward as true partners, each benefiting from, without harming, the other in life's dance.

SECTION FOUR

CREATORS OF THE FUTURE

Lend me the stone strength of the past and I will lend you the wings of the future, for I have them.

—Robinson Jeffers, *To the Rock That Will Be a Cornerstone*

Nearly 80 million children will be born worldwide in *each* year of the first decade of the new millennium. Many of these children will live to see the human population double. Before these children die, a world economy that is presently struggling to support 6 billion people will need to support 14 billion. That is more people than the total number of humans who have lived on earth since *Homo sapiens* first evolved over 400,000 years ago.

Such growth will dramatically magnify some of our existing challenges: preventing the spread of infectious diseases and limiting our pollution of the environment. It will also intensify our competition for the earth's finite resources—for places to live, work, and play, for water, and for food. In the face of our increasing demand for resources, we will challenge the very biosphere on which we depend for life.

Our voyage into the world of microbes has provided us with a new and broadened perspective about our biosphere. We enter the 21st century at the very beginning of our partnership with these most extraordinary creatures. Until recently, most of our explorations of the microbial world have been merely as observers. Now we are learning to understand, enter into, and change the biochemical and genetic conversations that take place among the microbes.

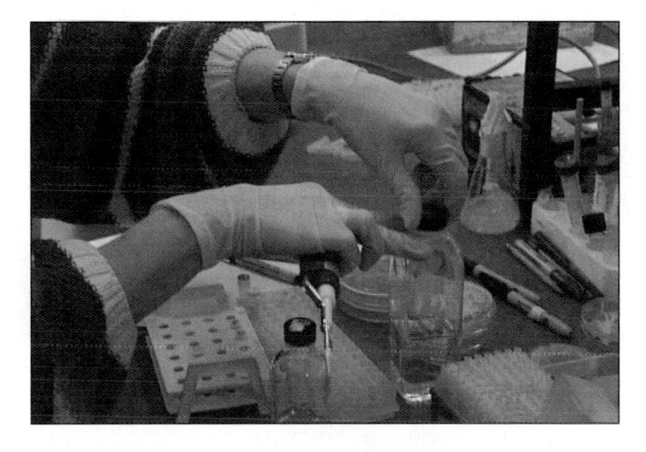

With microbes as partners, we are developing new ways to prevent and treat infectious diseases. We are helping microbes work more quickly to cycle our wastes back into raw materials and to remove toxic pollutants from our environment. We are using microbial tools to create plants that make more efficient use of space, water, and energy, promising at least a reduction in world hunger. We are learning to see microbes as miniature factories, offering new production tools to increase our industrial efficiency.

Our ability to solve problems by enhancing our microbial partnerships will present us with unprecedented challenges and unintended consequences. If we create new ways to control infectious diseases, we will need to decide who will benefit from the new treatment options and who will pay. If we eliminate infectious diseases, we will need to decide how to cope with the potential for even higher growth rates in our population. We will need to decide who will pay for cleaning up our old messes, who will grow the new foods, and who will own the microbial factories.

Ultimately, we will face an even larger dilemma. Given our growing ability to intervene in the complex system of checks and balances that has sustained life on this planet, we will have to decide whether and when we exceed our limits as a species by using technology to arrogate to ourselves an increasing share of the earth's resources.

Our partnerships with microbes offer new tools to meet difficult challenges—overcoming antibiotic resistance, cleaning up our environment, and feeding the growing world population. Each challenge presents us with difficult questions and the possibility of unintended consequences, including the ultimate challenge—the possibility that our new tools may be turned against us.

Combating Antibiotic Resistance

Alexander Fleming, working in his laboratory, inadvertently left the cover off a dish in which he was growing pathogenic bacteria. He noticed spots where specks of mold were growing and the bacteria were not. The growth-inhibiting mold turned out to be *Penicillium notatum*, the common bread mold, and the compound it produced was penicillin.

Penicillium

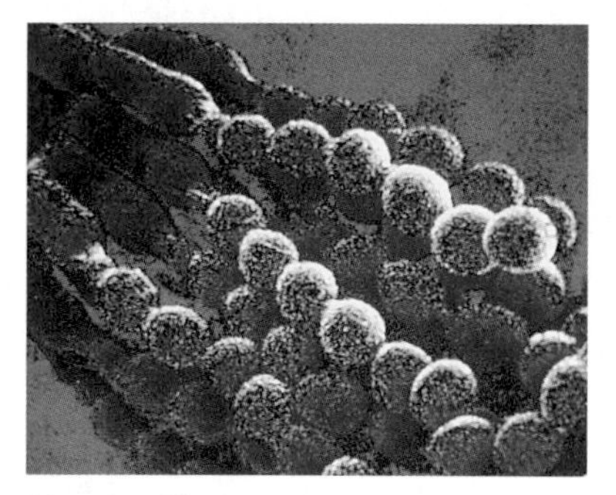

Identity: Fungus
Residence: Soil and decaying material
Favorite pastime: Decomposing
Activities: A member of The Producers, a particular strain of *Penicillium* manufactures the antibiotic penicillin for the pharmaceutical industry; others simply turn your bread and oranges green.

Throughout human history, a small number of microbes have been among our most dangerous enemies. As recently as the early 20th century, infectious diseases were the leading cause of death, accounting for millions of lives lost worldwide every year. Five of the top ten leading causes of death in 1900 were infectious diseases. By the 1990s only influenza and pneumonia remained in the top ten in the industrialized countries.

We owe this remarkable shift to a variety of factors. Chief among them is the discovery of antibiotics. You may think of antibiotics as weapons that humans created to fight pathogens—microbial disease producers. Actually, scientists have just adapted to our own needs a weapon that microbes themselves have been refining across their billions of years of evolutionary history.

Most antibiotics are produced by microbes. They interfere in one way or another with the cellular machinery of other microbes. Scientists are not entirely certain what their purpose is in nature, but it is likely that they serve as a means for keeping encroaching neighbors at bay.

In fact, most of the antibiotics in use today are modified chemically after a microbe has produced them. Modification may make an antibiotic effective against a broader range of microbial pathogens or change other properties, such as how well it is absorbed from the digestive tract.

Today we have over 100 antibiotics at our command. Most of them are targeted at bacteria. Bacteria are relatively easy marks in part because most carry out their life functions independent of their host's cells. Scientists have discovered many antibiotics that interfere only with bacterial growth and consequently don't cause significant damage to human cells.

Bacteria are also relatively easy marks because they have unique cell structures. One of these, called a cell wall, provides an exceptional target for antibiotic molecules. A cell wall is a tough casing that encloses the cell like the shell of a nut. Since human cells don't have walls, antibiotics that disrupt a bacterium's cell wall usually have no effect on our own cells. This distinction is the key to the effectiveness of the significant array of antibiotics called penicillins and cephalosporins.

Viruses, fungi and yeasts, and protozoa like *Giardia* are more challenging targets. Viruses live inside their host's cells, and fungal, yeast, and protozoan cells are very similar to human cells. Antibiotics that affect them are likely to affect our own cells as well.

Back to the Future

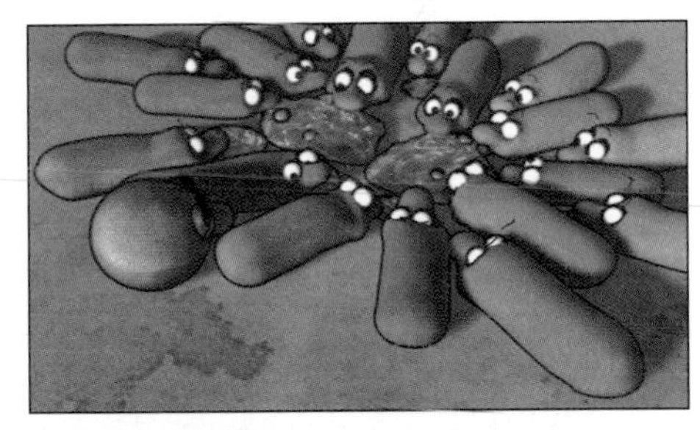

An alien microbe approaches other microbes as they enjoy a meal.

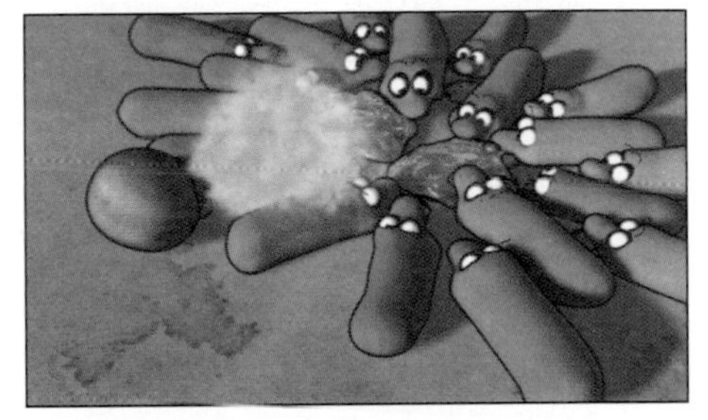

The alien secretes its own brand of antibiotic, dispersing its competitors . . .

. . . thereby gaining access to their food and space.

With Alexander Fleming's discovery of penicillin in 1928 and Florey and Chain's demonstration of its therapeutic potential in 1940, the balance between man and microbe seemed to change. Though antibiotics are still not universally available, their influence has nonetheless been felt around the globe. As each new antibiotic came into use, infectious diseases seemed more and more remote—a relic of the past.

But another shift in the equation has begun to temper our optimism. Within months after penicillin first came into use, strains of resistant microbes began to appear. The miracle drug had no effect on them. The problem seemed small at the time, but over the years it has grown to alarming significance. With each new antibiotic, the pattern has been repeated.

Now the promise of antibiotics seems to be fading. We are faced today with a rising tide of antibiotic-resistant microbes that cause serious disease. In some rare cases, the microbes are untouchable by modern medicine, resistant to every single antibiotic in our armamentarium. Patients infected with these resistant microbes are dying, much as people did before Fleming brought us his historic discovery. Some fear we may be returning to an era that we thought was past—an era without the benefit of modern antibiotics.

How we arrived at this point in history is an ominous story with valuable lessons for all of us.

The Microbes Fight Back

Since bacteria are frequent targets of antibiotics, they have developed many techniques for evading them—techniques that we call antibiotic resistance.

Bacteria can neutralize an antibiotic by chemically modifying the antibiotic's molecules. They accomplish this by producing enzymes that either attach a new chemical group onto the antibiotic molecule or break some of the chemical bonds that hold the molecule together. Take penicillin. This powerful antibiotic interferes with the enzymes used to build the bacterial cell wall by binding to them and making them inactive. However, many bacteria now possess other enzymes, called penicillinases, that break apart the penicillin molecule, rendering it useless.

Genes that provide the instructions for such enzymes are commonly found on the small bits of DNA that bacteria pass to other bacteria by horizontal gene transfer. The lucky cell receiving such a gene only has to carry around a little extra genetic material in order to gain the resistance trait.

Bacteria can also alter the parts of their own cell machinery that the antibiotic targets. This takes a bit more ingenuity, since the bacteria must accomplish the change without affecting their own normal cellular activities. In the case of penicillin, for example, some bacteria even go so far as to change the way they make their cell wall. After the change, penicillin can no longer level its harmful effects against the now resistant microbes.

Bacteria can even develop devices for keeping an antibiotic from ever reaching its target. They achieve this feat either by altering the outside of their cells so that the antibiotic can no longer get inside to do its work, or by pumping the antibiotic out of the cell interior as fast as it comes in. Both of these devices require very complex changes in the bacteria, a testament to how far a microbe can evolve in its own defense.

Similar devices exist among the fungal and protozoan pathogens and among the viruses. Scientists have discovered only a few antibiotics that can be used successfully against these microbial pathogens, but where antibiotics exist, these microbes can defend themselves as resourcefully as the bacteria do.

1 A thriving colony of microbial pathogens encounters a regiment of penicillin molecules.

4 Occasionally a mutation occurs, rendering a cell resistant to an antibiotic attack.

7 When the mutant divides, it makes exact copies of itself.

2 When the penicillin molecule attaches to a bacterium, the cell can no longer build its wall . . .

3 . . . and consequently dies.

5 When such a resistant mutant encounters penicillin . . .

6 . . . it remains untouched in an otherwise decimated colony of susceptible cells.

8 These survivors further divide . . .

9 . . . creating an entire colony of resistant cells.

Encouraging resistance

Antibiotics fed to animals, sprayed on fruit trees, or taken to cure the common cold encourage the emergence of antibiotic-resistant pathogenic bacteria.

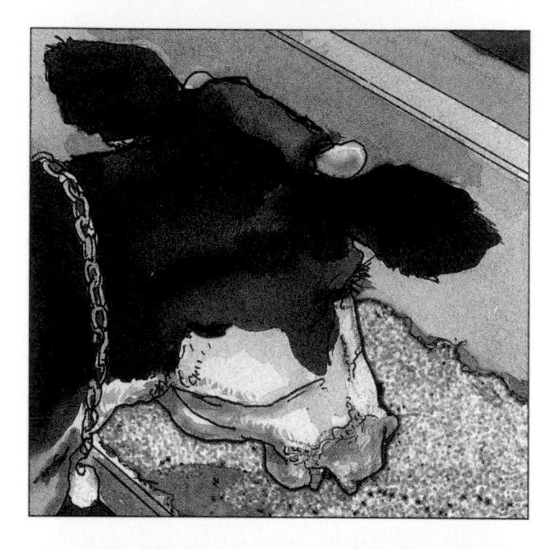

A Predictable Predicament

Knowing how remarkably adept microbes are at capitalizing on the processes driving evolution, we should have been better prepared for the emergence of antibiotic resistance. The basic mechanisms for evading an antibiotic's effects were developed in nature long before we started using antibiotics for our own purposes.

We have contributed to the problem of antibiotic resistance in a variety of ways. We produce and consume an enormous amount of antibiotics annually. In 1954, about two million pounds of antibiotics were produced in the United States. By 1998, the U.S. alone was producing over 50 million pounds of antibiotics annually.

Sometimes the well-intentioned use of antibiotics to treat human infections can be misguided. The Centers for Disease Control and Prevention in the U.S. estimates that only about one-half of the antibiotics given to people to treat infection are needed. Physicians, often under pressure from their patients, prescribe antibiotics even when they know they are not effective.

Take the common cold. Thus far, researchers have not found an antiviral agent for the viruses that cause the infection. Yet physicians admit to prescribing antibiotics to treat the common cold—often just to keep their patients happy.

Even if antibiotics are prescribed appropriately, they are not always used properly. As soon as people begin to feel better, or experience unpleasant side effects from the antibiotic, they all too commonly quit taking it. Although the symptoms may be gone, the offending microbe may not be, especially the members of the invaders that were a little less susceptible to the antibiotic. Once unencumbered from the antibiotic assault, the less susceptible microbes are free to reproduce, thus increasing their numbers in the microbial population at large.

The inappropriate use of antibiotics is more extensive in countries where antibiotics can be purchased over the counter, thereby increasing the chances that the antibiotic will be used when it is not necessary or in suboptimal doses. More antibiotic use leads to more selective pressure, which leads to increasing numbers of resistant microbes.

Surprisingly, only about half of the antibiotic supply is used to treat human infections. An additional 40% is used in animal husbandry. For reasons that are unclear, the addition of an antibiotic to animal feed seems to promote livestock growth. The amount available to any individual animal through its feed is not sufficient to treat an infection, but it is more than sufficient to select for antibiotic-resistant strains of bacteria. Farmers use the remaining 10% to control bacterial infections of economically valuable plants in agriculture. They may spray acres of fruit trees, for example, with solutions of antibiotics before the fruit is picked. This leaves behind a residue, both on the fruit and in the environment, which selects for resistant bacteria.

We can acquire the resistant bacteria and their genes by eating plants and animals. Although the microbes may not cause disease in us, their resistance-encoding genes can be transferred to our own normal bacterial flora and to common human pathogens.

All of these practices translate into circumstances that are ideal for the emergence of antibiotic-resistant bacteria. Couple the large quantities of antibiotic used with the large reservoirs of genes encoding antibiotic resistance in the environment and the microbial ability to transfer genes rapidly throughout surviving populations, and it's a wonder we still have effective antibiotics left at all.

Reversing Reversals

Scientists have learned that the number of antibiotic-resistant microbes in the environment drops relatively rapidly when the selective pressure of antibiotics in the environment is reduced. We can, therefore, adopt strategies that allow us to use antibiotics more wisely—to treat disease while limiting the amount of unnecessary antibiotic poured into our environment.

Crowding facilitates the spread of disease among people and the consequent likelihood that resistant organisms will be passed to others and thrive.

Some of the strategic elements involve changes that can begin today. We can, for example, personally avoid unnecessary use of antibiotics. Pressuring a physician for a prescription to treat a cold or the flu increases the likelihood that he or she will prescribe a useless antibiotic.

Other strategies will follow from technological developments that lie in the future. For instance, certain kinds of antibiotics pose a particular problem. Broad-spectrum antibiotics are drugs that affect many different kinds of bacteria simultaneously, providing the selective pressure for many different kinds of microbes to develop resistance. Such antibiotics are usually prescribed when physicians are unsure what microbe is causing a disease, or once known, whether the microbe is susceptible to a more specific antibiotic. The use of rapid laboratory tests that identify pathogens in minutes rather than days can help physicians to choose much more specific antibiotics. Such rapid tests, like the ones available for detecting the group A streptococcus that causes "strep throat," can help to improve antibiotic choices.

Neither better tests nor improved use of antibiotics gets us out of the hot water we are currently in, however. This dilemma will become more pressing as populations grow and crowding increases the transmission of contagious diseases. It is increasingly important that we find new strategies for controlling and treating infectious diseases.

Scientists hunt for novel antibiotic-producing microbes in soil samples near Chernobyl, Ukraine, the site of a major nuclear accident in 1986.

Microbe Safaris

Yuri Gleba, a Ukrainian scientist, examines the branch of a pine tree whose genes have been altered by radiation.

One way of securing our health and well-being is to look to the microbial community itself for help. Scientists speculate that there are many novel compounds yet to be discovered among the microbial world's inhabitants. By applying the knowledge they have gained about microbial interactions, along with advanced technology, researchers should be able to discover them.

To better appreciate future approaches to antibiotic discovery, it helps to understand how antibiotic prospecting was done in the past. Researchers would collect samples of soil from the environment. In the laboratory they would distribute the samples into little plastic dishes filled with nutrients, in the hopes of coaxing some of the microbes present in the sample into growing and reproducing.

The laboratory was far from a natural setting for the microbes, as they were now separated from their communities and living on strange food. Once they were reproducing in the nutrient-filled dishes, however, each kind of microbe could be tested to determine whether it produced compounds with antimicrobial properties.

There are scientists who are still pursuing this traditional approach. They are doing their hunting, however, in novel settings. A case in point are the microbe-hunting safaris that are going on in the highly radioactive environment that surrounds nuclear reactor #4 in Chernobyl in Ukraine.

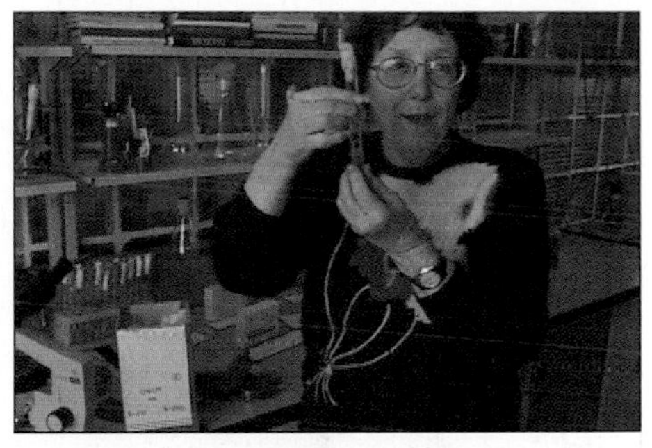

Ukrainian scientist Elena Kiprianova examines samples of microbes grown from the soil around the site of the Chernobyl disaster.

On April 26, 1986, Chernobyl was the site of the world's most disastrous nuclear accident, which killed 15,000 people and left the area within an 18-mile radius heavily contaminated with radioactive material. The radiation left behind had a toxic effect on everything, including the microbes that survived the explosion and live in the soil surrounding the reactor. Those microbes that remain have had to adapt to a drastically altered environment. Their communities have changed.

Scientists like Yuri Gleba from the Institute of Microbiology in Kiev and Jennie Hunter-Cevera from the University of California, Berkeley, come to the area around Chernobyl to collect samples. They hope to discover microbes that produce novel compounds as a consequence of adaptation to this highly radioactive environment. Their expectation is that some of the compounds may be antibiotics.

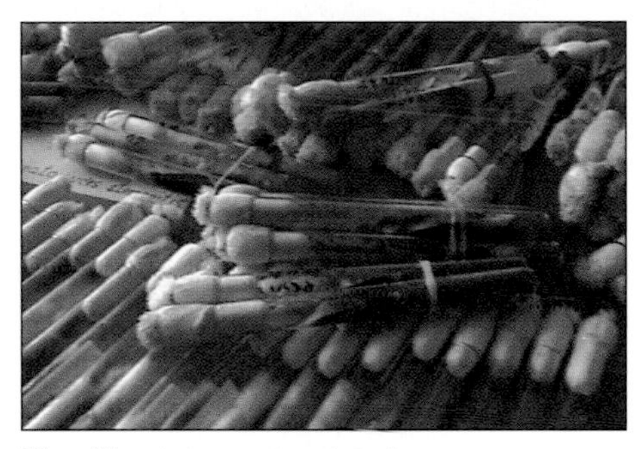

The Ukrainian scientists have successfully grown several thousand species of microbes from around the world.

"What the combined Ukrainian/U.S. program offers is the chance to link two very different sciences—the classic and the modern—and actually have more knowledge in combining those two sciences."

—Jennie Hunter-Cevera

Scientists at the University of British Columbia collect samples of soil from a park in Vancouver.

Gene Safaris

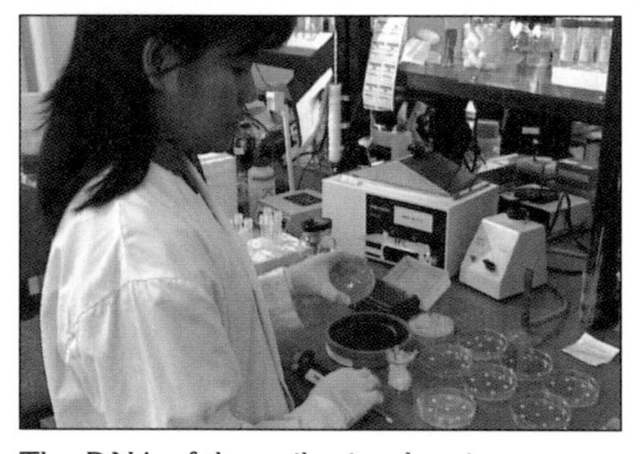

The DNA of the soil microbes is extracted, cloned, and then inserted into microbes that can be grown readily in the laboratory, in the hope that their new genes will yield new antibiotics.

The obvious limitation to this kind of search for antibiotics in nature is that scientists can only test the microbes that will reproduce in the laboratory—in vitro. Scientists now know that this represents only about 0.1% of all the existing kinds of microbes, and there is every reason to think that among the 99.9% of the microbes left to be grown, some will produce antibiotics. The challenge is to find the microbial compounds without first having to grow the microbes in the laboratory.

Julian Davies at the University of British Columbia has spent much of his professional career battling antibiotic resistance, and he has now mounted a new effort. He and his fellow scientists are using a strategy that capitalizes on the ability to extract DNA from microbes living in their natural habitats. They are using this strategy to look for the genes that instruct microbes that can't be grown in vitro how to make antibiotics. Once these genes are extracted, they can be inserted into bacteria that are easily grown in the laboratory. Researchers can then probe the genetically modified bacteria for new compounds with antibiotic-like activity. Even a tablespoon of soil from your back yard holds 5,000 to 10,000 different kinds of microbes, most of which have not been grown in the laboratory. Julian Davies and others are betting that some will produce antibiotics that haven't been discovered yet.

A few laboratories around the world are using this technique to screen millions of these genetically transformed surrogates to find new antibiotics. The process is slow and tedious; the time from discovery to application for any antibiotic is often 10 to 12 years. The cost of drug development is high, regardless of the approach, but many scientists believe the gene safaris will at least lead us to many new and useful antibiotics.

From the Dirt Beneath Your Feet

Getting from a bag full of soil collected from the forest floor to a new antibiotic by extracting the DNA and cloning it into a microbial surrogate might seem complicated. In practice, it's easier than it sounds.

The first step is to sift out the large particles of debris in the dirt. Then the dirt is mixed with special chemicals in a blender—creating a sort of a mud milkshake. The mixture contains all of the five to ten thousand kinds of microbes that were present in the soil sample.

The scientists are interested in the microbes' DNA, in particular the genes they carry that may code for new antibiotics. So the soil microbes are next mixed with a chemical that breaks the cells open, releasing each cell's DNA into the liquid. From here, the DNA can be easily separated from all the other material in the liquid.

The microbes' extremely long strands of DNA are far too large to work with in the laboratory, so the DNA must next be cut into small pieces so it can be efficiently manipulated. This is accomplished by using another microbial product—a protein called a restriction enzyme. The restriction enzyme works like a pair of molecular-sized scissors, cutting the DNA into small strands.

Now there are millions of short strands of DNA, any of which may contain the genetic information for making a new antibiotic. The next trick is to get the DNA from the soil microbes into a surrogate—a bacterium like *Streptomyces* that can be easily grown in the laboratory. The surrogate can then take over the role of the soil microbe, producing the compounds that are encoded on the transferred DNA.

Recombinant DNA technology provides a way for scientists to transfer the genetic material into the bacterial surrogate. Bacteria are in the habit of passing around their DNA using a carrier called a plasmid. A plasmid is a short strand of DNA, connected into a circle, that can move genes in and out of cells. Scientists use other enzymes to insert the small DNA fragments from the soil microbes into the plasmid circles in a process that is the molecular equivalent of film splicing. The newly programmed plasmids with their spliced-in genes can now be easily moved into surrogate *Streptomyces* cells and incorporated into their genetic library. Scientists call these surrogate cells recombinants, and the process they use after the cells have reproduced is called cloning.

The recombinant cells grow to large numbers in the laboratory, making the products that are encoded on their spliced DNA. Scientists can now look for products with antibiotic activity without ever having to coax the original soil microbes into reproducing in vitro.

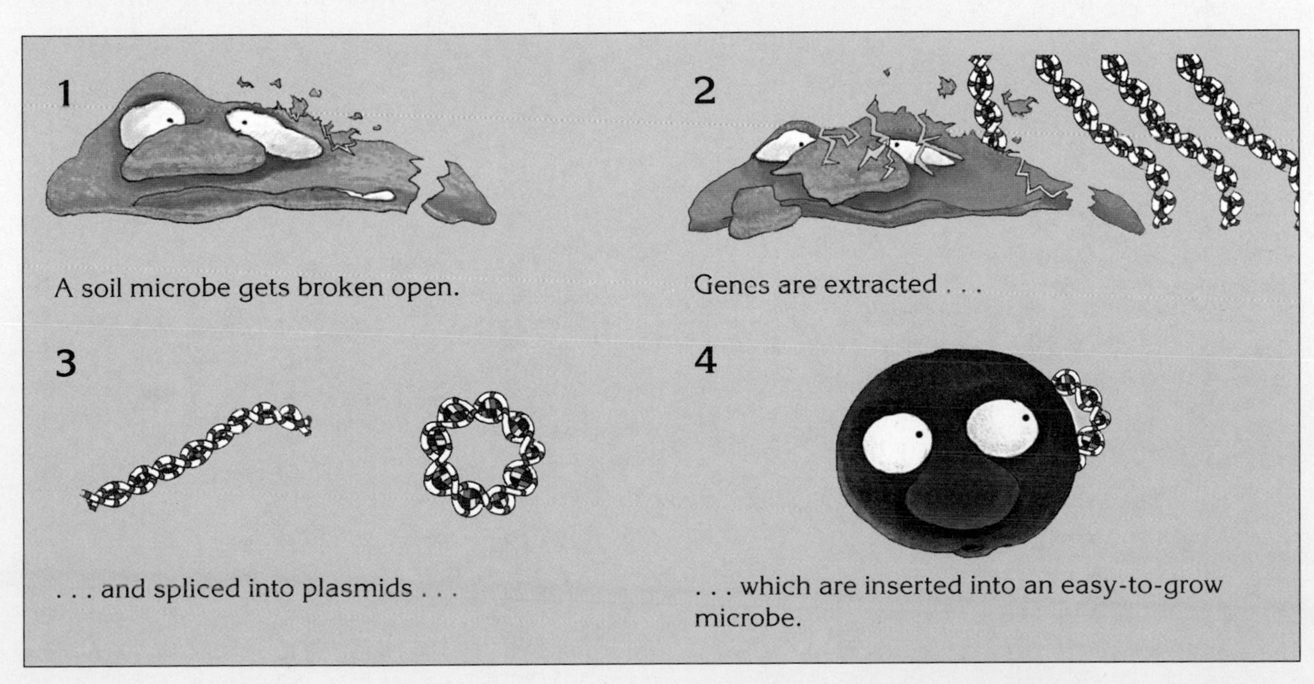

1 A soil microbe gets broken open.

2 Genes are extracted . . .

3 . . . and spliced into plasmids . . .

4 . . . which are inserted into an easy-to-grow microbe.

Blocking Microbial Conversations

Even if researchers are successful in discovering new antibiotics, physicians will still be left using the same approach to treat infections that they do now—administering an antibiotic that will slow the growth of, or kill, an invading microbe. If antibiotics are used properly, in a controlled fashion, and when prescribed only by physicians, they will have a much longer useful life. History, however, has taught us that in the long run this approach is destined to fail.

Ultimately, the solution for stemming the tide of antibiotic resistance may depend on a different approach. What if scientists could find a way to block a pathogenic microbe's ability to produce damage, while allowing it to reproduce normally? Would that make the development of antibiotic resistance less likely?

Researchers know that many disease-producing microbes have developed a chemical dialog—a chemical conversation with the cells of their host. In our case, the conversation may convince our cells to allow the offending microbes to cross our boundaries, it may cause our immune cells to delay their normal killing action, or it may cause certain of our cells to change the way they run some of their internal machinery. Designing compounds that

Quorum sensing

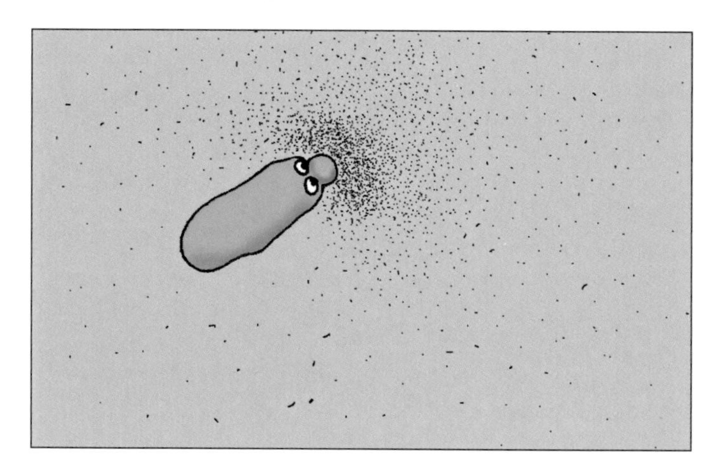

Microbes secrete chemical signals . . .

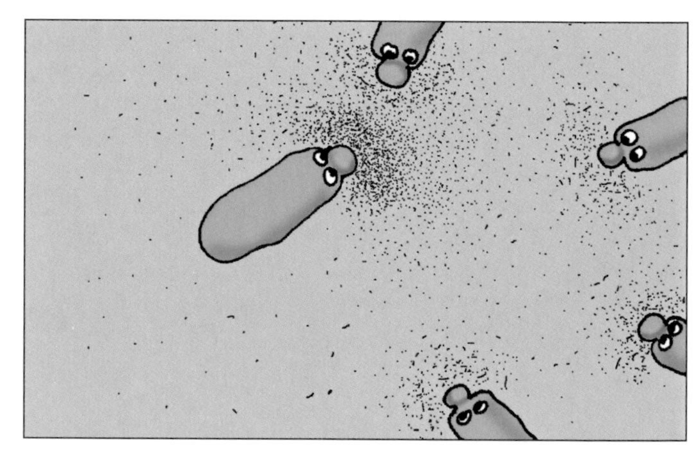

. . . that fellow microbes recognize.

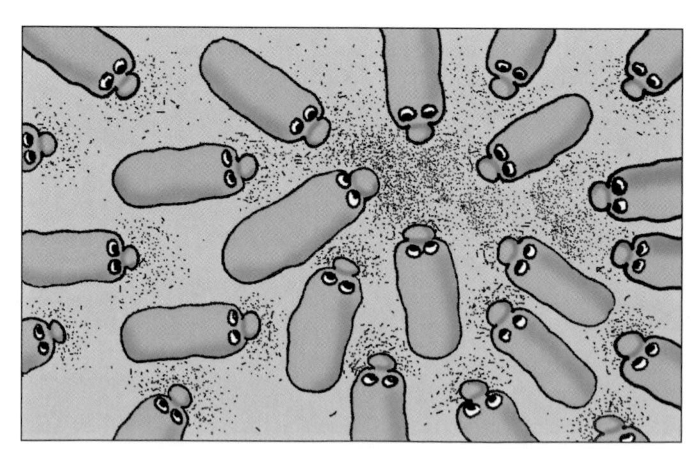

When their numbers reach a critical mass . . .

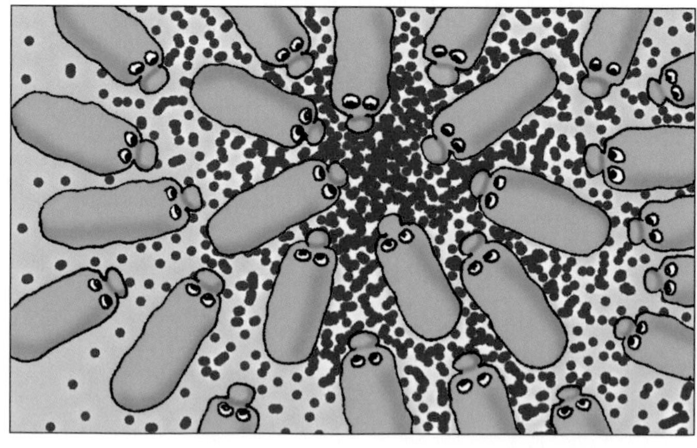

. . . they release new, and sometimes damaging, chemicals.

block the disease-causing conversations might protect us against the microbe's harmful effects while not presenting a serious disadvantage to the microbes' urge to reproduce.

An even more intriguing approach is based on the recognition that microbes are social creatures. They employ chemical signals to tell other microbes of their type that they are in the vicinity. Scientists call this phenomenon "quorum sensing" because the chemical noise tells a specific microbe when there is a quorum of like cells present. In the case of pathogens, "hearing" the right signal triggers the microbe to manufacture an array of virulence factors, most of which are not produced until a sufficient number of cells are present to launch a successful invasion.

Consider the human pathogen *Staphylococcus aureus.* This microbe causes a wide range of disease, from skin infections and boils to bloodstream infections and toxic shock syndrome. *S. aureus* damages our cells by using an extensive set of toxins and other chemicals, but the microbe only produces these virulence factors when it senses a quorum. If researchers could find a way to make *S. aureus* "deaf" to the message that signals the presence of a quorum, they might be able to block its production of virulence factors, staving off serious damage to us while allowing the microbe to continue reproducing normally.

Streptomyces

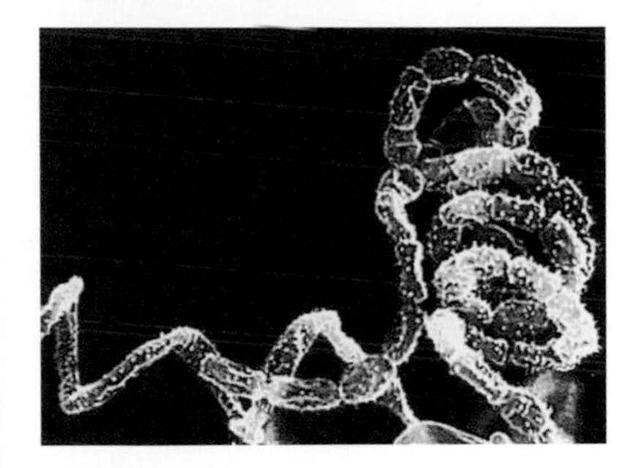

Identity: Bacterium
Residence: Soil
Favorite pastime: Producing geosmins, the compounds that give soil its characteristic smell
Activities: A member of The Producers, the *Streptomyces* are prodigious synthesizers of antimicrobials, producing over 500 different chemical compounds with antibiotic activity.

Some DNA with Your Chips?

The genetic revolution is yielding better tools for diagnosing infections. Traditional diagnostic tests often require days to weeks to obtain a result. New genetic tests allow physicians to identify specific pathogens in hours. Better diagnostics mean a better approach to treatment, reducing some of the factors that contribute to the emergence of antibiotic resistance among pathogens.

Researchers have learned that each microbe has certain unique sequences of nucleotides within its DNA. Scientists make these unique sequences, sometimes called molecular probes, for any given microbe and attach an array of the probes, each representing a different microbe, to the surface of a silicon chip. Laboratory workers then wash a specially prepared sample collected from a patient across the chip's surface. The DNA of a pathogen present in the sample from the patient will react with the portion of the array associated with that specific pathogen. Laboratory workers scan the surface of the chip with a laser to determine the exact location of the matched DNA strands. Using a map of the chip's surface, a computer can identify the probe and, hence, the microbe.

If the molecular probe is for *Salmonella,* for example, then physicians know that *Salmonella* is present in the sample the laboratory has tested. If the molecular probe detects a gene conveying resistance to a specific antibiotic, then physicians know to choose an alternative antibiotic. If all of this can be accomplished within a few minutes after a clinical specimen has been collected and tested, the physician has a much better chance of selecting just the right therapy for the infection at the earliest possible time.

Cleaning House

With all the benefits of industrialization have come the unintended social costs of industrial pollution. In some circumstances, we have so polluted our environment that tracts of land have become uninhabitable and water undrinkable.

Building nuclear weapons generated large quantities of radioactive waste that must now be stored. Microbes resistant to such high levels of radioactivity may play a role in future cleanups.

The pollution problem in the United States alone is staggering. Twenty million Americans live within four miles of one of the government-designated "superfund" sites—the sites where the most dangerous contamination problems exist. Over 4 million people live within one mile of a superfund site. The latest estimates of how much it's going to cost for the labor, disposal areas, and construction equipment to clean up just the sites that we know of—the worst sites—is $1.7 trillion.

Scientists around the world are working on solutions based on a new understanding of the microbial mechanisms for keeping life running—that one creature's waste is often another's banquet. Microbes with the right appetites can play a big role in cleaning up our past and helping us prevent contamination in the future.

A Cold War Legacy

Scientists at Savannah River in Georgia are counting on such mechanisms. They are utilizing the microbes that live in their neighborhood, using methods that long predate human existence. The uniqueness of their approach to environmental cleanup lies in the ability to encourage local microbes to speed up the natural process of environmental recycling.

The government defense plant at Savannah River, Georgia, is one of the United States superfund sites. Sites like these are heavily contaminated and pose a threat to the environment.

The government built the Savannah River Site in the early 1950s to produce the materials—primarily tritium and plutonium-239—used to build nuclear weapons. Five reactors produced these elements by irradiating target materials with neutrons. The newly minted nuclear substances had to be cleaned up before they could be used to make weapons, so workers next moved the material to one of two separation plants at Savannah River. There they used a chemical process to separate the desired nuclear material from waste. The separation process required highly toxic solvents including trichloroethylene and tetrachloroethylene, each sometimes called TCE.

Workers transferred the used solvents to holding tanks. Periodically, they released the chlorinated solvents through tile-lined pipes buried 20 feet underground and running to a large basin. There the solvents were allowed to evaporate.

Unfortunately, the pipes were leaky. An area one mile square, running along the path of the pipes, became heavily contaminated with TCE.

Although this scheme seems unthinkable in retrospect, scientists and engineers in the 1950s thought that evaporation was the safest and most effective means of disposal for these solvents. Chemical manufacturers routinely issued standards of practice that called for disposal by evaporation. It was years before we realized that these chemicals would seep down and contaminate our water supplies.

At Savannah River, the water table lies 100 feet underground, beneath the old pipes. As the TCE slowly seeped out of the pipes and down through the soil, it entered the groundwater, potentially contaminating drinking water wells, streams, and rivers that the groundwater fed. From there, the TCE could—and most likely did—enter the food web.

Chlorinated solvents like TCE are among the most difficult of contaminants. They are stable in the environment for many years, they are toxic in incredibly small amounts, and they can cause cancer. The acceptable level in the environment for TCE is less than five parts per billion. In other words, a single five-gallon container can contaminate up to one billion gallons of water.

Environmental scientist Terry Hazen examines gauges at the Savannah River, Georgia, superfund project site.

The conventional approach would have been to pump the water up out of the ground and treat it to remove the toxic material. All you accomplish by this approach, however, is to move the toxic material to another location, and transport itself is not without risk. A team of scientists led by Terry Hazen, a veteran ecologist, turned to the microbial world for answers, and the answer they got from the microbes astounded even the most imaginative among them.

Caution, Microbes at Work

The microbial players the scientists used belong to a group of bacteria called methanotrophs. Methanotrophs occur naturally in the soil and water, and they share a unique characteristic—they all use methane as a source of food.

Methanotrophs contain an enzyme, called methane monooxygenase, which is a key to converting the methane in the environment into their cell material. By combining methane molecules with oxygen, the bacteria use what they need for energy and release the remaining part of the molecules to the atmosphere as carbon dioxide.

Methane is the major constituent in the natural gas that heats our homes and fuels our stoves. It abounds throughout the natural environment. And where there is methane, there are methanotrophs happily existing off of the gas.

Hazen and the scientists and engineers working at the Savannah River site knew that methane monooxygenase will also degrade over 250 other compounds. Among these are some of the world's most toxic chemicals, including TCE. The investigators knew that methanotrophs were likely to be present in the contaminated soil and already slowly degrading the TCE. They also knew that, left to their own devices, the microbes would take decades to complete the process.

The question was how to increase the microbes' activity to levels that would degrade the contaminants in a shorter time.

Methanotroph

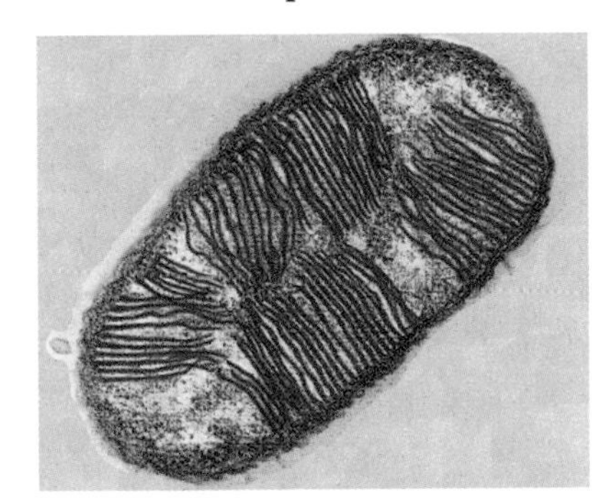

Identity: Bacterium
Residence: In water and soil, wherever methane is present
Favorite pastime: Hanging out around decomposers
Activities: A member of The Degraders, methanotrophs use the methane released through decomposition as food, converting it into cell material and carbon dioxide.

A methanotroph lives on methane in the soil but is capable of eating some 250 other organic compounds in a pinch.

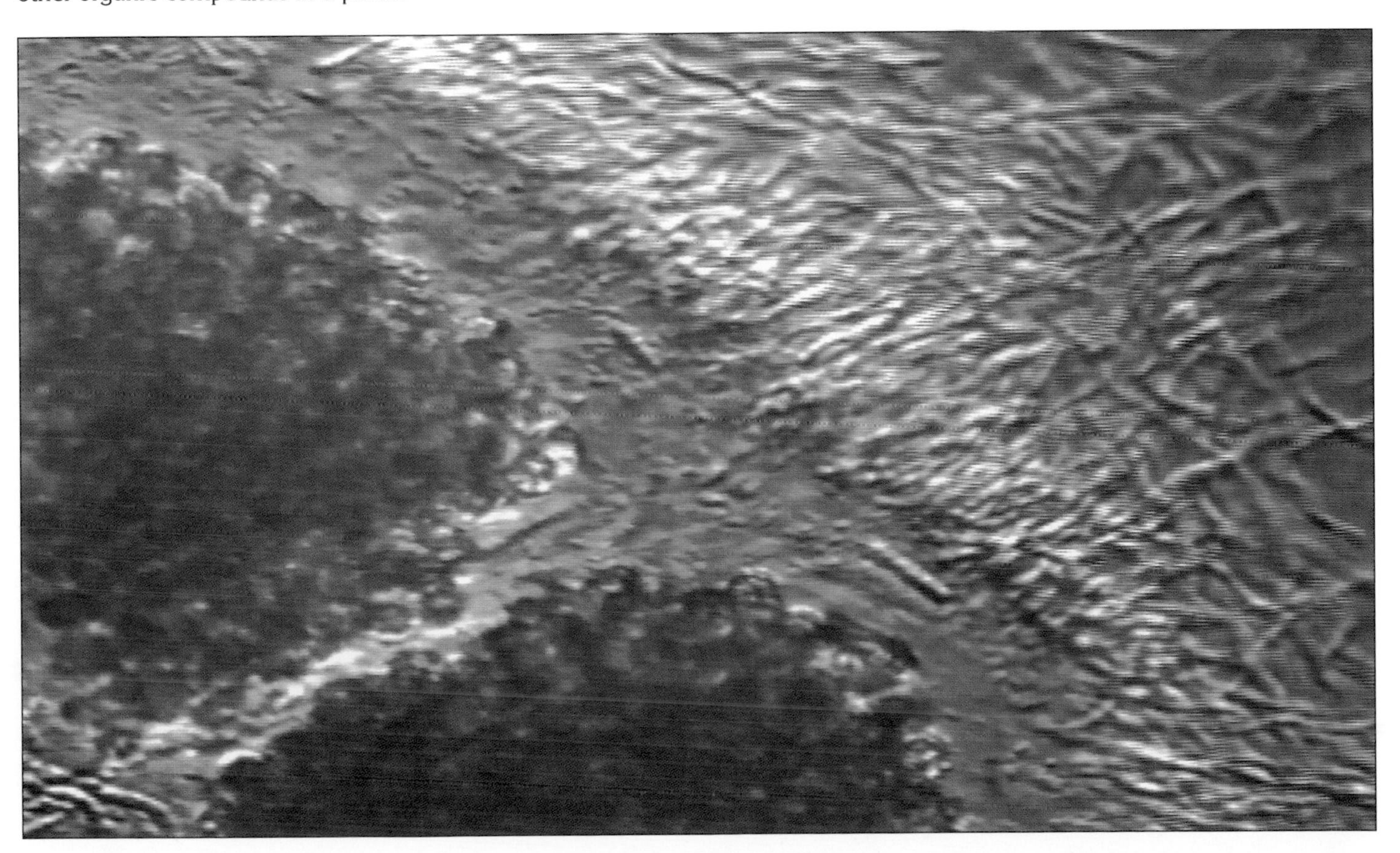

The answer came from a marriage of biology with the tools of the petroleum industry. The scientists reasoned that if they could increase the concentration of methane in the soil, the resident methanotrophs would have plenty of food. They might then replicate faster, increasing their numbers dramatically. If subsequently the methane concentration were reduced, the methanotrophs might be inclined to substitute other compounds they could use, like TCE.

A fine way to increase the concentration of soil methane was to run a horizontal pipeline right under the contaminated groundwater. The engineers commandeered some of the horizontal well-drilling equipment used by the petroleum industry to drill for oil. They laid a line 100 feet down along the route of the old pipes—the ones used to move the contaminated material to the evaporation basin—and they began to pump natural gas and oxygen through the lines.

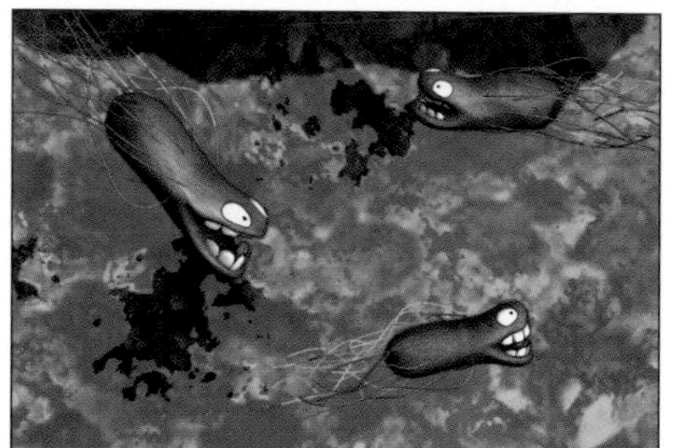

In metabolizing methane and other chemicals, methanotrophs release carbon dioxide and salt.

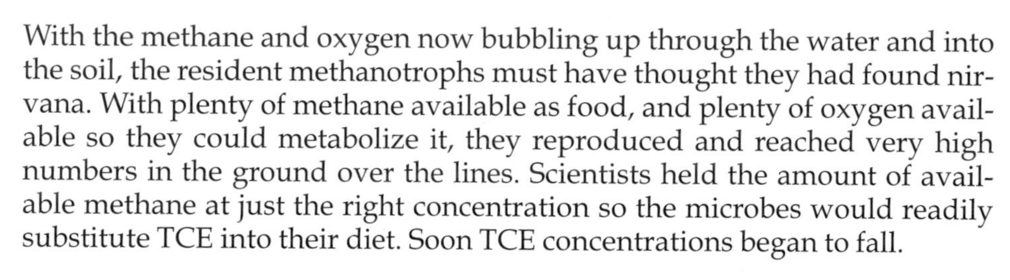

With the methane and oxygen now bubbling up through the water and into the soil, the resident methanotrophs must have thought they had found nirvana. With plenty of methane available as food, and plenty of oxygen available so they could metabolize it, they reproduced and reached very high numbers in the ground over the lines. Scientists held the amount of available methane at just the right concentration so the microbes would readily substitute TCE into their diet. Soon TCE concentrations began to fall.

Within a period of a few months—not years, and not decades—the concentration of TCE had dropped to undetectable levels. All that remained were the by-products of its consumption: carbon dioxide and salt. Once the microbes' work was completed, the engineers turned off the natural gas. Without the steady supply of food, the methanotroph population declined back to its original levels, and this particular area was restored to much the way it was before the Savannah River Site came into being.

The cost of this operation was about 45% of what the cleanup would have demanded with conventional technology, and the process took far less time. And the microbes degraded the toxic chemicals naturally, leaving behind completely benign substances. The scientists had found a way to harness one of the oldest of nature's solutions—a process called bioremediation.

Methanogens to the Rescue

Scientists at Savannah River, Georgia, drove long pipes below the area of pollution . . .

. . . and pumped in methane gas, which bubbled up and allowed the natural population of methanotrophs in the soil to increase rapdily.

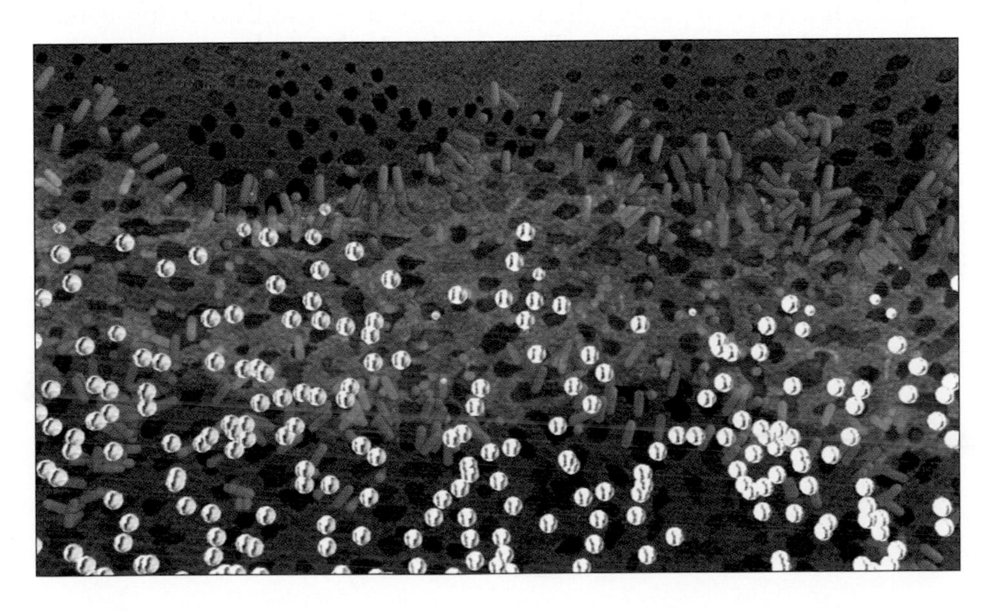

When the gas was turned down, the dense population of microbes was forced to turn to the pollutants as their food source.

Who Takes Out the Garbage Anyway?

Terry Hazen and his colleagues from the Savannah River Environmental Restoration Program are looking for applications for microbes in places closer to home. Take the Baker Road Landfill in Columbia County, Georgia. It's just across the road from the Savannah River Site, and it's one of 88 landfills in the state—otherwise known as the local dump, where household and industrial garbage ends up when it's thrown away.

In 1990, Georgia, along with most of the other states in the U.S., changed its rules on landfills. The Environmental Protection Agency recognized that landfills were a serious source of groundwater contamination. Water draining out of the vast masses of refuse often contained chemical contaminants that slowly filtered into the water supply. From 1990 on, all landfills had to be constructed with a liner underneath to trap the wastewater as it drains through the garbage. The contaminated water can then be removed and treated. Once the landfill is full, it must be capped and monitored as the material within slowly degrades over a period of 30 or more years. All of this makes for a very expensive process.

Hazen and his colleagues convinced the state of Georgia and the EPA to allow them to run a demonstration project. They argued that the Baker Road landfill was like a giant compost pile, only with potentially toxic waste material in it. Using composting strategies, it should be possible to speed up the degradation of the organic material, allowing it to compress more rapidly and thus take up less of the precious space allotted to it.

Composting is a purely microbial process. It relies on the consumption and digestion of organic waste (like vegetable, fruit, and paper trash) by a complex community of naturally occurring microbes—life's normal way of taking out the garbage. The microbes convert much of the trash into humus that can be added back to the soil. By piling our garbage up in giant heaps within the landfill, however, we deprive the

In landfills, the microbial decomposition of trash is slow due to lack of oxygen. Hazen and his colleagues stimulated the microbes' growth by turning over the materials and pumping in oxygen, making a huge compost pile. In short order the microbes converted the trash to simple organic compounds, carbon dioxide, and water.

microbes of sufficient oxygen and moisture to do their job efficiently. Consequently, it takes years to degrade the material.

Hazen and his colleagues turned one site within the landfill into a carefully monitored compost pile by pumping in oxygen and recycling the trapped water through the mass to keep the moisture content at the proper level. The primary microbial by-products—heat, water, and carbon dioxide—were monitored, and the material was stabilized so that it could be compressed within a matter of months. By reusing the contaminated water, they limited the amount of liquid waste that had to be treated at the end of the process.

So far, the experiment seems to be working. While the landfill composting experiment won't result in the same kind of usable soil additive that we expect from a backyard composter, it is helping reduce the amount of contaminated solid and liquid waste. And that's a good start.

"It's sort of interesting that the microbes . . . the smallest of God's creatures, have the greatest potential for curing the greatest problems."

—Terry Hazen

The microbes convert the trash to simple organic compounds, and the waste products—carbon dioxide and water—are released into the air.

A High-Tech Legacy

Northern Vermont, in the northeastern United States, attracts millions of visitors every year. They come to enjoy the rolling green hills of summer, the brilliant colors of fall, and the skiing and beauty of winter.

Tom Watson was one of many who came to Vermont, fell in love with the country, and made it a second home. But Watson was different. He brought more than his love of the outdoors to northern Vermont. Tom Watson was Chairman and CEO of IBM, and he brought a major manufacturing plant to a small town named Essex Junction. The plant has become one of the world's largest manufacturers of computer chips—the tiny computer parts that power the information age.

The IBM manufacturing plant in Essex Junction is a bit like the iron and steel foundries of the past. The manufacturing facility has brought employment and prosperity to the local citizens, but the chemical character of the manufacturing process has had an environmental impact on the soil and groundwater beneath the IBM site.

Semiconductor manufacturing is a chemically intensive process, requiring various solvents, acids, bases, and gases to build the tiny electrical circuits on the silicon wafer. The chemicals are used to draw and etch the semiconductor circuit patterns in the silicon wafer and clean the wafers to the exacting standards required to prevent damage to the tiny circuits.

Like many companies in the 1960s and 1970s, IBM stored and transported solvents and solvent waste in underground systems for safety reasons. As

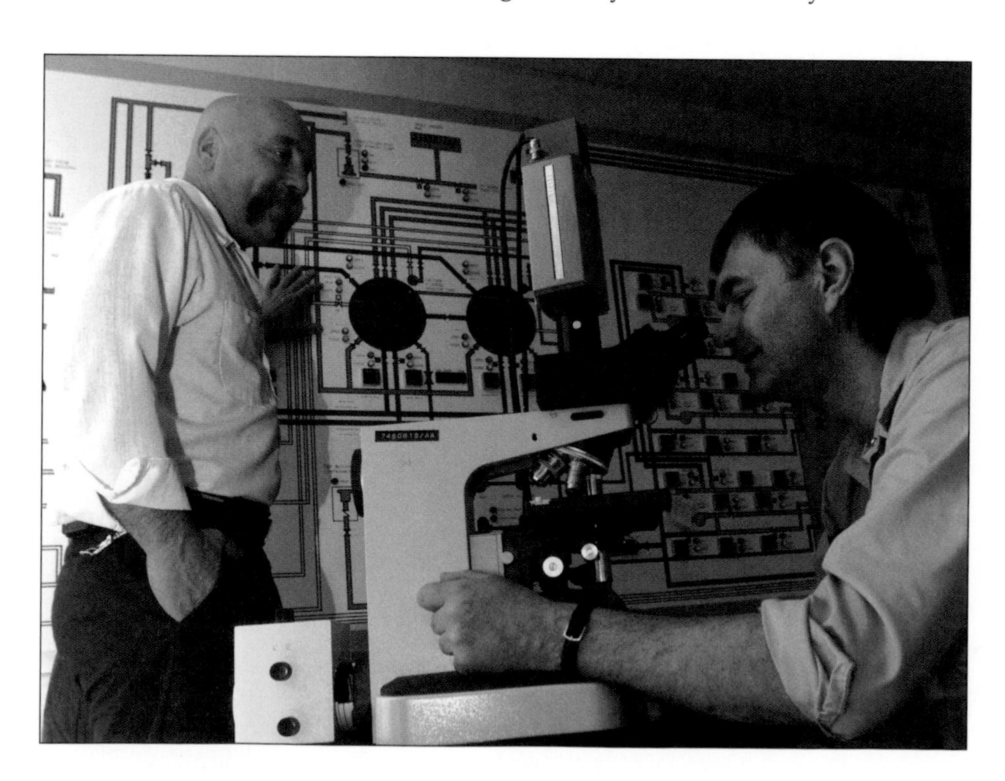

Managing toxic wastes with the help of microbes is an increasingly attractive industrial practice. Here, workers at the IBM chip manufacturing plant in Vermont use sophisticated monitoring devices to control for leaks and use microbes to turn their solvents into benign substances.

the solvents were considered to be highly volatile, spills of material were not cleaned up because they would "evaporate away." Hindsight has shown that these practices resulted in solvents being left in the soils and groundwater beneath the chemical piping and handling areas.

Since discovering the problem in 1979, IBM has installed above-ground, spill-contained, leak-detected chemical piping and tank systems to prevent the release of chemicals to the environment. IBM has also negotiated a permit with the U.S. Environmental Protection Agency and the State of Vermont to undertake a cleanup program at the facility. Groundwater is pumped and treated to remove contaminants. Air is drawn through the soils, collecting the solvent materials for removal in a filtration unit.

IBM has discovered an unlikely ally in its efforts to clean the solvents from the groundwater. Naturally occurring microbes have evolved the ability to "eat" the solvents, reducing them to carbon dioxide and water. In many areas, this microbial action has accelerated the removal of the solvents in a manner similar to the Savannah River site. The microbial digestion has been so successful that IBM has begun to use the naturally occurring microbes to remove solvents in its treatment unit.

Past contamination aside, IBM must also continue to manage the wastes from its ongoing operations. Both processes—cleanup and waste management—are expensive, and the cost passes on to employees, shareholders, and consumers. IBM had to find a more environmentally friendly manufacturing process to protect their future.

First, IBM engineers changed the production process so that isopropyl alcohol (IPA) could be used to clean the chips rather than the more toxic solvents used before. This helped, but the Essex Junction plant generates about 40,000 gallons of IPA waste each year, waste that can no longer just be dumped on the ground.

Initially, IBM trucked the IPA wastes to a local facility for incineration. This was costly, and still created a pollution problem: releasing greenhouse gases into the atmosphere.

In 1996, IBM began processing the IPA wastes in a microbe-based reactor, where countless invisible new partners provide an alternative to burning the spent solvent. This microbial feeding frenzy continues there today, consuming the IPA and releasing a benign waste product back into the reactor.

After the biological process is completed, IBM treats the now detoxified waste in an industrial wastewater treatment plant and ultimately releases essentially pure water into the adjacent Winooski River. Microbes have provided IBM with a solution that saves the manufacturing operation almost $40,000 annually and results in reducing the contaminant to a safe and biologically usable compound. The latter is priceless.

IBM is still left with its legacy from earlier contamination. The corporation has shouldered this responsibility, but how they will achieve the cleanup remains an open question.

Strain T1

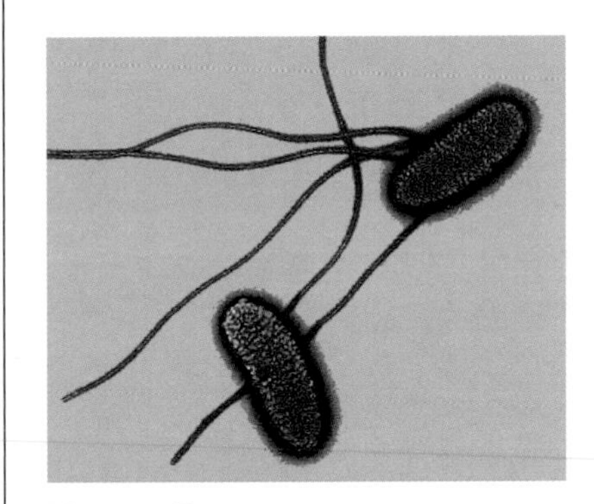

Identity: Bacterium
Residence: In the soil
Favorite pastime: Seeking places where oxygen doesn't exist
Activities: A member of The Degraders, this bacterial strain can consume and digest toluene, a chemical that is toxic for people, but provides a good lunch for T1.

Inventing Microbes

The need to clean up our environment sometimes extends beyond the natural abilities of the resident microbes. This is especially true when there are multiple toxic chemicals present, or when the chemicals include those designed specifically to withstand the effects of microbial degradation. In the majority of these situations, we are left dependent on traditional methods of environmental remediation—removal and storage.

The microbial world may yet come to our rescue in this most difficult area. Given enough evolutionary time, they most certainly would. But evolutionary change takes a long time. Consequently, scientists have been attempting to preempt evolution by combining the desirable traits of individual microbes into a single "super-bug."

In the early 1970s, a group of scientists led by Ananda Chakrabarty created one of the first super-bugs. They introduced genes from several different microbes into a single microbe, giving it the ability to degrade toxic compounds found in petroleum. The enhanced microbe could ultimately thrive on crude oil alone, degrading some of its components to usable material and energy. This super-bug thus offered an attractive alternative to skimming and absorbing spilled oil.

The oil-degrading microbe has never been released into the environment. It did, however, play an important role in establishing the biotechnology industry. The U.S. Patent Office granted Chakrabarty the first patent ever for the construction and use of a genetically engineered bacterium. This established a precedent allowing biotechnology companies to protect their "inventions" in the same way chemical and pharmaceutical companies had in the past.

We are entering an era where we are accomplishing with biologic methods tasks formerly done by machines.

Since that time scientists around the world have been taking advantage of the tools that microbes themselves use. Strains designed to solve the most complex environmental cleanups can be a reality. By moving the genes for altering organic solvents into a microbe that can survive extremely high levels of radiation, for example, we would have a microbial partner ready to help in the management of solvent waste contaminated with radioactivity.

The release of genetically engineered microbes into the environment, however, raises concerns among thoughtful citizens. So far, such releases have been severely restricted, limiting our ability to test super-bugs like Chakrabarty's in the wild.

The *Exxon Valdez* oil spill in Alaska highlighted the need for better ways to clean up petroleum and the many toxic chemicals that it contains. The spill was the first time that nutrient enrichment was proven to speed the naturally occurring microbial degradation of oil.

We have no precedents for this technology, but we have had some unfortunate experiences with the introduction of exotic plants and animals into native environments. The introductions of gypsy moths, zebra mussels, African killer bees, and Asian milfoil have all had unintended and devastating consequences.

Whether genetically engineered microbes bear any similarity to these examples is questionable. But it is important to remember that we live in a world of ecosystems, not just organisms. Since ecosystems are webs of tightly linked relationships, changing one component is likely to change others. The other changes are often unpredictable. And the greater the scale of the change, the greater the number of unforeseen consequences.

Feeding the World

The world's increasing ability to grow subsistence crops has bolstered food supplies over the past two decades, yet undernutrition remains a serious problem in many developing countries because of population expansion. We will need to increase global food production by another 50% in the next 50 years just to stay even with global population growth.

Solving this problem will depend on many factors—expanding cropland and irrigation, minimizing soil degradation, and improving farming practices and the efficiency of water use. It will also depend on improving the crops that we now use.

Genetic manipulation to improve plants is as old as farming itself. We have traditionally created plants with desirable traits through selective breeding. If we discovered a plant that produced larger fruit or more seeds, we would propagate that strain in preference to others. If we wanted a plant with better drought resistance, we would crossbreed it with another plant that had the desired traits.

Selective breeding is a slow and tedious process, however. Microbes offer us tools to speed incorporation of valuable characteristics into existing plants and, in some cases, provide valuable traits from the microbes' own genetic pool.

Saccharomyces

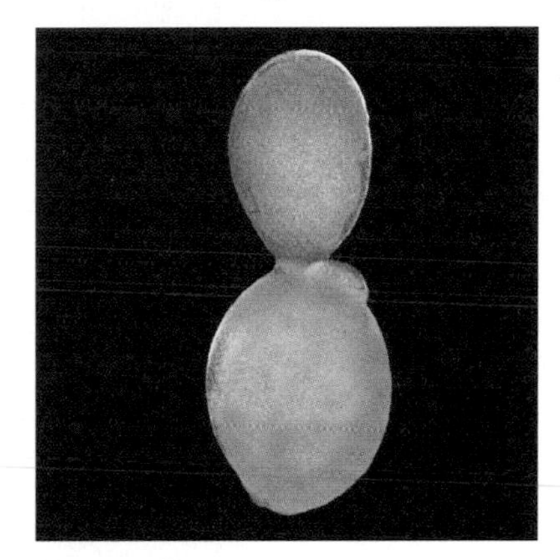

Identity: Fungus
Residence: On fruits and flowers, where sugar is present
Favorite pastime: Making gas and alcohol
Activities: Among the oldest members of The Producers, *Saccharomyces* strains are best known as baker's and brewer's yeasts; they have participated in the making of bread, beer, and wine for thousands of years.

The Microbial Genetic Engineer

Scientists have turned to a clever genetic engineer, *Agrobacterium tumefaciens,* as a way of getting beneficial genes into plants. This bacterium infects plants, generating a disease called crown gall. *A. tumefaciens* does this by inserting some of its DNA into the plant cell it is infecting. The genes so transmitted are responsible for the disease, which results in a tumor-like growth on the plant.

Scientists learned that they could eliminate the bacterium's disease-producing capabilities without altering its ability to insert DNA, and hence new genes, into a plant cell. This de-fanged bacterium has become a useful partner for inserting all sorts of new genes into plants.

Once *A. tumefaciens* has successfully inserted new genes into individual plant cells, each cell can be grown into a whole plant. Every cell in the plant contains the new DNA and passes it on to its progeny. That means that seeds or cuttings from the plant all contain the new DNA, and all their progeny do as well.

A virus-resistant strain of cassava for Zimbabwe is such a plant.

The cassava root is a mainstay in the diet of over 500 million Africans.

Living with Nature's Challenges

Zimbabwe is in sub-Saharan Africa. The country typifies the present challenges posed by growing populations in developing countries. Over 70% of Zimbabwe's population rely on agriculture for their subsistence. They grow crops to eat and, when production is good, to sell. Most of Zimbabwe's farmers, however, live in areas that are very marginal for farming. The soil is poor and the weather is unpredictable. Droughts are common, and crop failure is more the rule than not. That means little or nothing to sell and often not enough to eat.

The main subsistence crop in Zimbabwe, as well as for 500 million other Africans, is the cassava. Cassava grows as a green, leafy plant rising 5 to 6 feet above the ground. The business part of the cassava, however, is its roots. The underground roots swell as they store the starch produced by the green plant above ground, developing into large, potato-like tubers. It's this part of the plant that provides a ready source of nutrition to the farmers and their families. A single cassava root can feed a family for several days.

Cassava is in many ways ideally suited as a major food source in sub-Saharan Africa. Its primary virtue is drought resistance. When other crops like maize and wheat have dried up and died, the cassava plant will still be standing and producing its nutritious tubers.

But there is a problem with growing cassava—it is susceptible to a devastating infectious disease. The infectious agent is African cassava mosaic virus. When the cassava mosaic virus takes over a cassava plant, it infects cells throughout the plant. The virus's replication causes the plant's leaves to become shriveled and spotted.

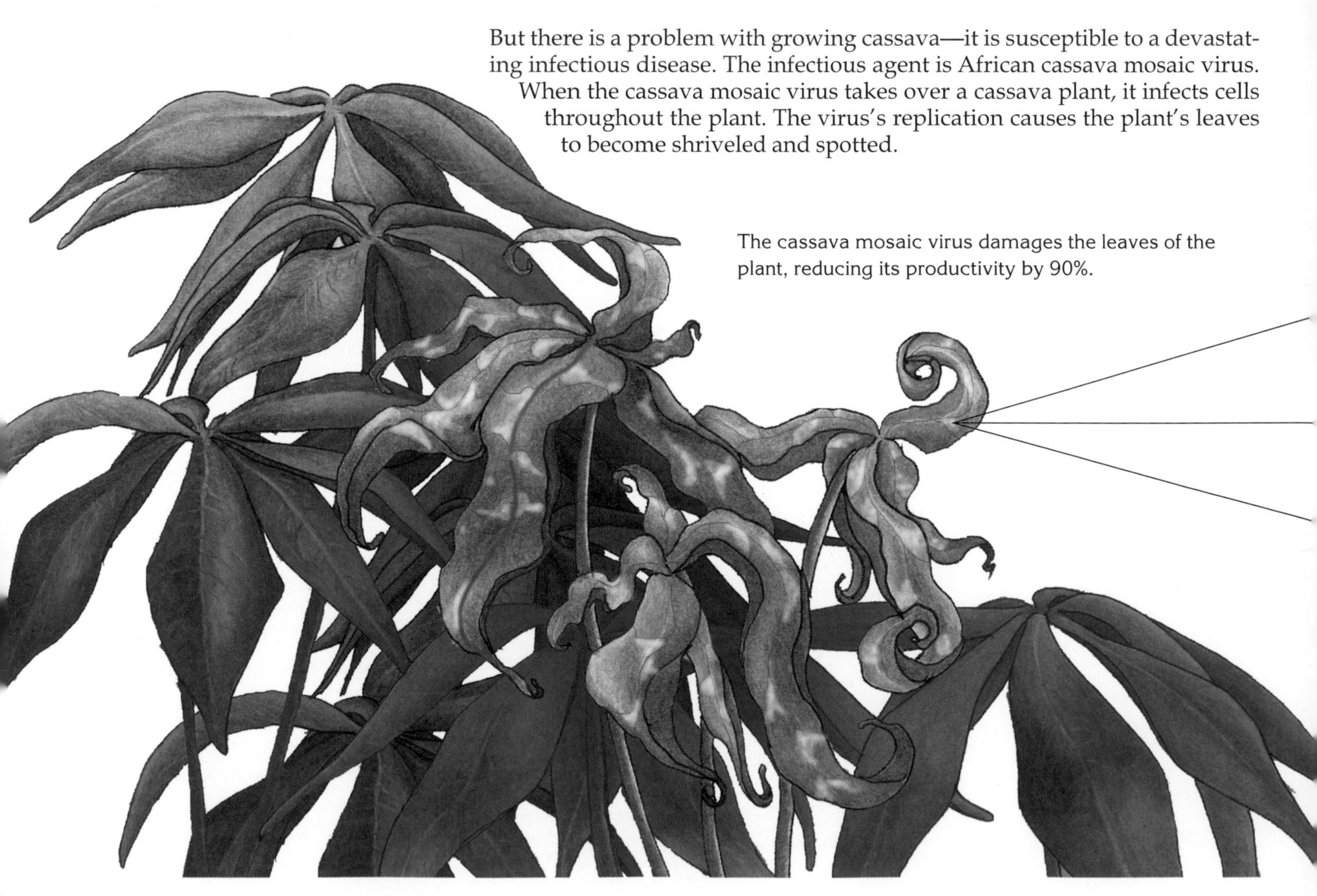

The cassava mosaic virus damages the leaves of the plant, reducing its productivity by 90%.

With damaged leaves, the cassava plant can't convert the sun's energy into usable material—in this case the starch that it stores in its roots—as efficiently. When the virus is at its worst, the plant's productivity may decrease as much as 90%. A farmer with an infected field will get only 3 to 6 tons of cassava per hectare rather than the 30 tons normally expected.

Like many viruses, the African cassava mosaic virus is hard to stop. Its primary means of spread depends on another common inhabitant of the cassava fields in Africa, the white fly. When the fly feeds on an infected cassava plant, it sucks in both its nutrient-rich sap and, incidentally, the virus. The fly carries the virus along to the next plant, injecting it into the plant's stem when it sucks up its next meal. Once the virus gets a foothold, it can be spread rapidly by this unsuspecting carrier.

One simple way to stop the virus is to stop the fly. The villagers could control the fly with pesticides, but the chemicals are expensive and can easily contaminate another scarce resource—their drinking water.

Moreover, the virus has another unsuspecting assistant—the villagers themselves. The farmers propagate the cassava by transferring cuttings from the plant's stem. If the plant is infected with the mosaic virus, transferring a cutting to a friend's farm also transfers the virus. Even without the white fly, the virus can spread from field to field. There is a great need to find a way to stop the virus—one that doesn't rely on chemical pesticides.

How the virus spreads

Transported through grafting from one diseased plant to another

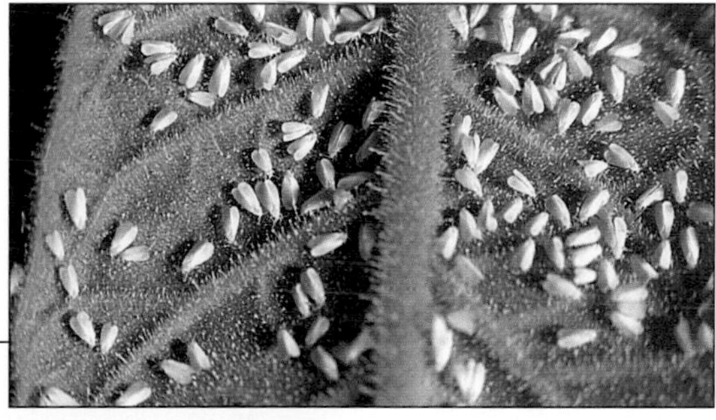

Carried from plant to plant by the African white fly

Passed via diseased roots from neighbor to neighbor

Cassava mosaic virus

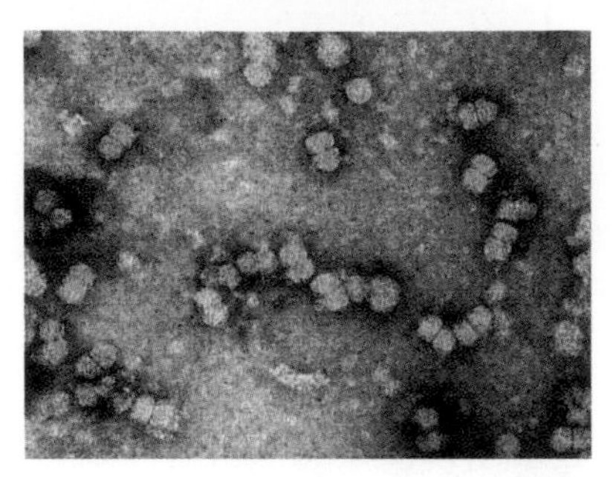

Identity: Virus
Residence: The cells of the cassava plant's leaves
Favorite pastime: Going aerial with its white fly vector
Activities: A member of The Pests, this virus infects an important food plant in Africa, crippling the plant's capacity to produce its large potato-like tubers.

Borrowing Nature's Solutions

Victor Masona, a young Zimbabwean scientist at the International Laboratory for Tropical Agricultural Biotechnology in La Jolla, California, and his colleagues turned to one of nature's own strategies to help them create cassava plants resistant to the attack of the virus.

Scientists working in other laboratories had learned that plants genetically engineered to produce a particular virus product called a coat protein are resistant to infection from that virus. The coat protein signals the commandeered host machinery that the last step in viral replication is complete, so the virus is fooled into believing that it has accomplished its reproductive mission. Masona and his colleagues hypothesized that they could create a cassava plant resistant to the mosaic virus if they could insert the gene for the virus's coat protein into the plant.

Masona turned to the bacterial genetic engineer *Agrobacterium tumefaciens* to test his hypothesis. He inserted the viral gene for its coat protein into *A. tumefaciens.* He then used the microbe to transmit the new DNA to his cassava plant cells. Once the gene had been successfully inserted into the bacterium, nature took its course.

Victor Masona, a Zimbabwean biologist, instructs the villagers about infected leaves.

The genetically engineered cassava plants proved to be a success in the laboratory. The scientists' hypothesis was proven: the plants that could make the mosaic virus's coat protein were indeed resistant to infection by the virus. They had created a new cassava plant, one that held great promise for the villagers who depend upon it in Africa.

Yet a major step still had to be taken. In order to demonstrate that the genetically engineered cassava would perform as expected, it had to be returned to its natural environment—Zimbabwe. The now virus-resistant strain of cassava would have to grow in the dry and poor soil and produce its nutritious tubers just like the virus-sensitive plants. And this final step turned out to be at least as difficult as the genetic manipulation of the cassava.

The Zimbabwe government has not yet determined its position on the introduction of a major crop plant that has been genetically manipulated. Regulations must be written and put into place before Masona and his colleagues can give the virus-resistant cassava plant its last and most critical test in the field. In the meantime, Africa's hungry villagers hang in the balance.

Agrobacterium tumorifaciens

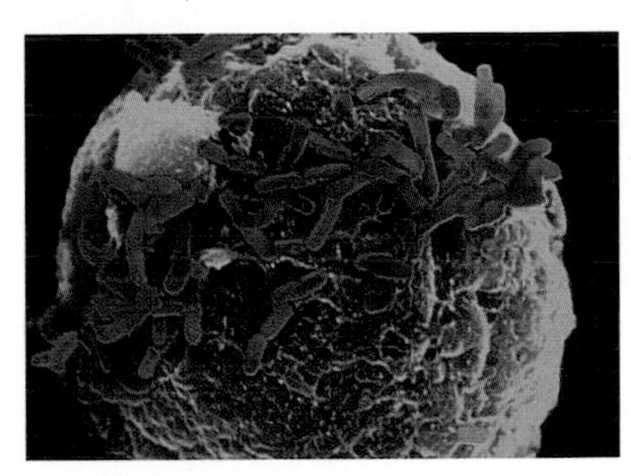

Identity: Bacterium
Residence: Plant tissues
Favorite pastime: Passing around genetic information
Activities: A member of The Gene Movers, *Agrobacterium* passes genes to plants that it infects, giving the plant crown gall disease, but providing us with a convenient way to transfer selected genes to plants.

"It can be very frustrating that the plants are sitting in the lab in California. I cannot proceed with the testing and introduction of genetically engineered cassava if the regulations that guide this sort of introduction are not in place."

—Victor Masona

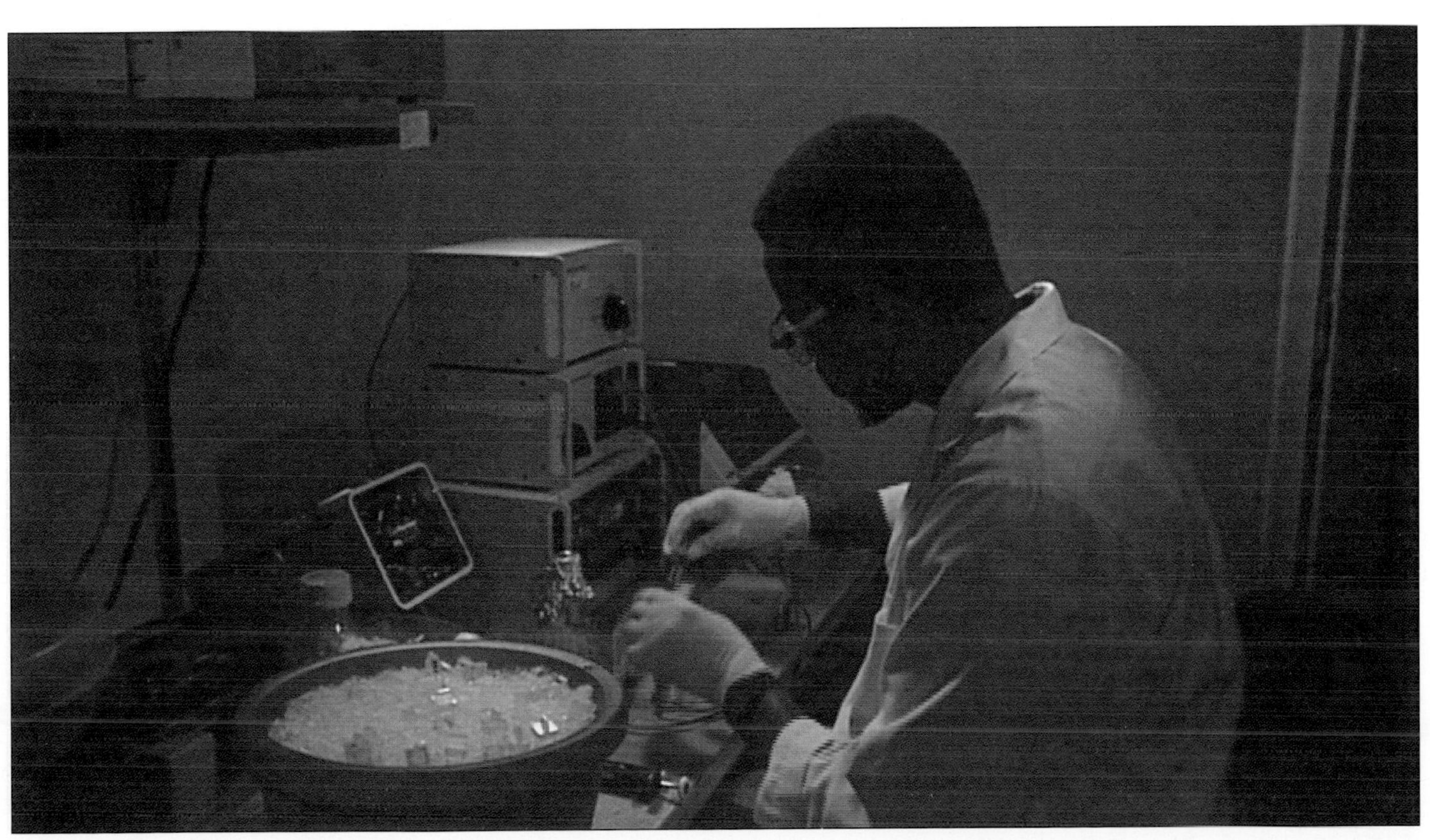

Making More and Better Food

Using *A. tumefaciens* along with other methods of transferring genes, scientists around the world have introduced valuable microbial traits into a wide variety of plants. Consider the plants that make their own pesticides, allowing them to fend off their own natural enemies without chemical pesticides. The New Leaf potato is just such a plant. Scientists gave this plant a bacterial gene for a toxin that is highly effective against one of its principal predators, the potato beetle. Since all of the potato's cells contain the gene, they all produce the toxin. Hence, the New Leaf is growing along happily in fields today even in the midst of ravenous hordes of beetles.

Microbes can also help make plants resistant to herbicides. Large-scale crop production depends on having an efficient mechanism for keeping weeds out of the fields, and growers have turned to herbicides to achieve that end. Unfortunately, herbicides are typically not very selective. They will kill the crop plants along with the weeds, so they have to be applied before the crop emerges.

In one effort to address this challenge, scientists have introduced into soybeans a bacterial gene that codes for resistance to a commonly used herbicide called glycophosphate. Since the modified soybeans aren't harmed by the herbicide, the grower can wait to make certain the chemical is really needed before applying it. The engineered soybeans are saving the grower dollars and the environment another insult.

Some plants have been given improved storage life, making it possible to get more food into the market and ultimately to the consumer. Take the Flavr Savr tomato. It contains a gene that interferes with the production of an enzyme that weakens the tomato's structure. With the enzyme's activity blocked, the tomatoes don't spoil as rapidly. They can be harvested later than normal tomatoes and still survive the trip from field to market.

A new genetically modified strain of rice promises improved nutrition for billions of people in developing countries who depend on rice as a staple. The modified rice grains are gold-colored because they contain a set of microbial genes coding for beta-carotene, a compound which humans readily convert to vitamin A. Vitamin A deficiency, which leaves people more susceptible to infectious diseases and at risk for blindness, affects over 400 million people worldwide. The golden rice also contains genes that allow the grains to accumulate extra iron in a form that humans can absorb. Iron deficiency is an even more important problem, affecting up to 3.7 billion people.

Victor's Vector

1 In nature, plants are subject to invasion . . .

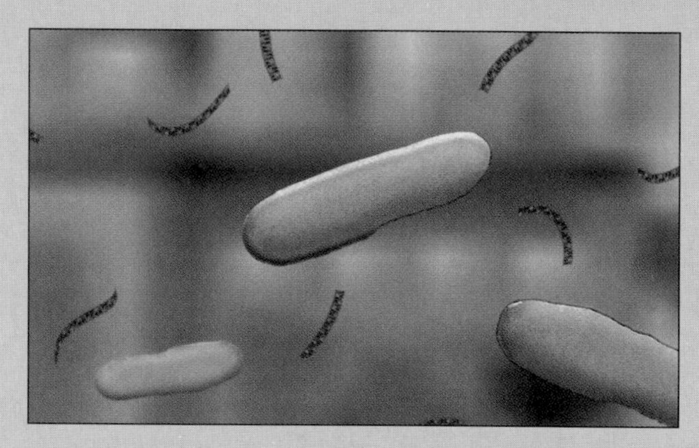

4 In the laboratory, *Agrobacterium* and the gene for the virus's coat protein are mixed together . . .

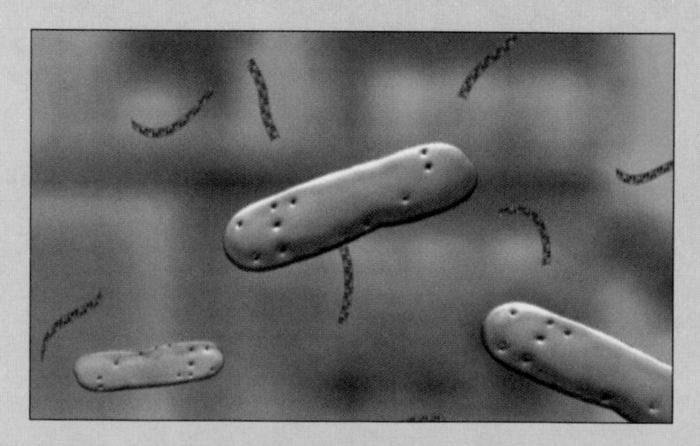

7 . . . allowing the virus's genes to enter their cells.

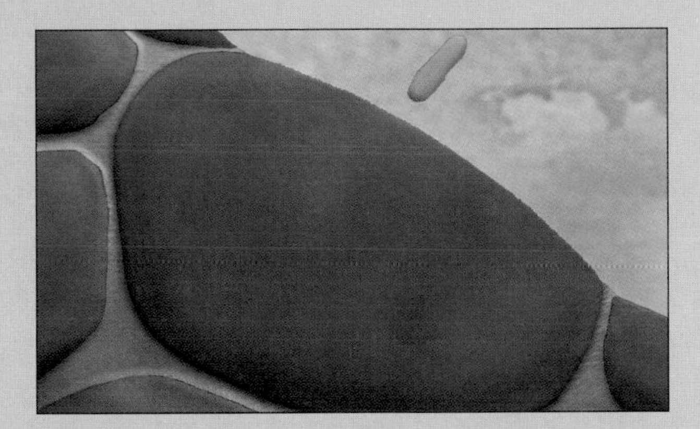

2 . . . by *Agrobacterium*, a bacterium which inserts its genes into the plant's cells . . .

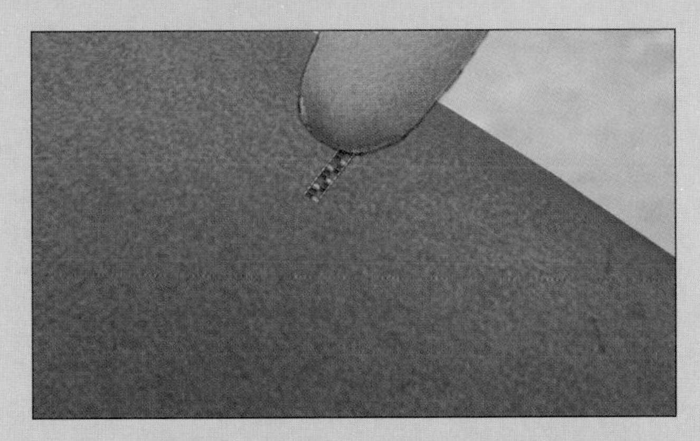

3 . . . but does little or no harm to the plant.

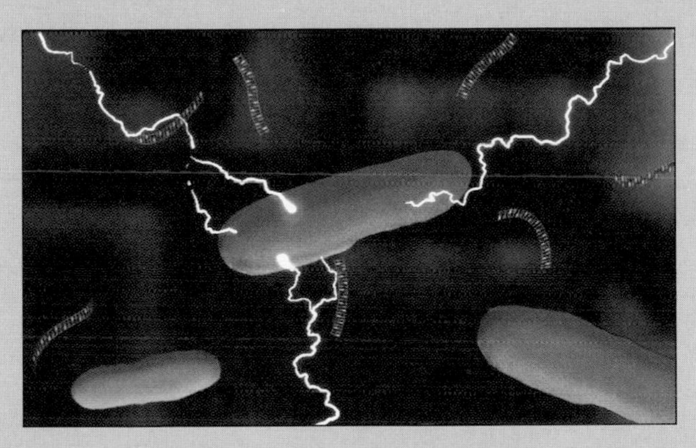

5 . . . given a jolt of electrical current . . .

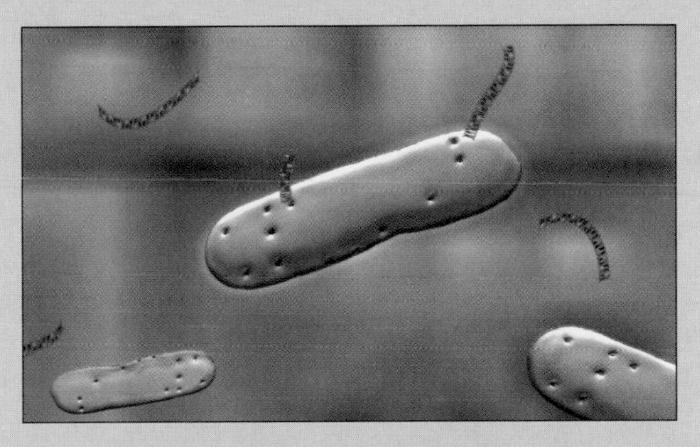

6 . . . which punches holes in the bacteria . . .

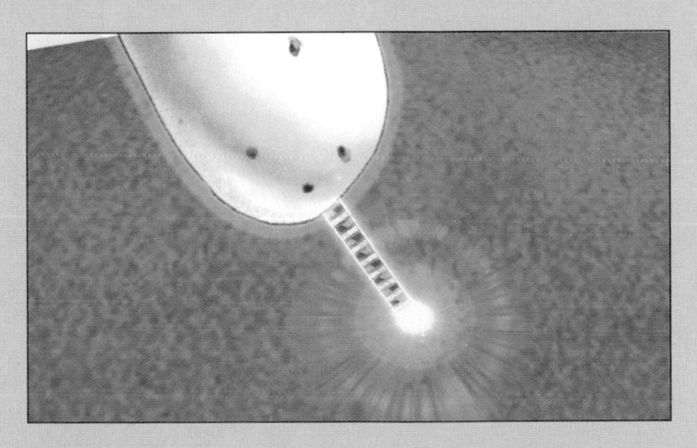

8 Now the genetically altered bacteria are allowed to infect cassava cells.

9 When the cells grow into a whole plant, its leaves are now resistant to infection by the virus. The plant has been "vaccinated"!

Turning Over a New Leaf

The Colorado potato beetle, long the scourge of potato growers in the U.S., can strip a potato plant clean of its leaves virtually overnight. Potato growers protect their plants from this voracious eater by spraying the plants with pesticides. This is no more appealing to the U.S. grower than it is to the African farmer. The practice is expensive and potentially contaminates the environment.

Enter the New Leaf. The New Leaf potato is different in one sense from all other potatoes: it produces its own pesticide. It didn't acquire this trait through natural selection, however. Scientists were able to introduce a gene from a bacterium called *Bacillus thuringiensis* into the potato. The gene codes for a protein called Bt toxin that is lethal for any unsuspecting potato beetle that might attempt to nibble on the New Leaf's new leaves. Hence, planting the New Leaf saves at least one dousing with a more traditional pesticide.

The New Leaf's introduction into the marketplace has engendered controversy, despite the plant's seemingly desirable characteristics (unless, of course, you're a potato beetle). Traditional potato growers in the U.S. have generally greeted it with open arms. Anything that can save them even one application of pesticide to acres and acres of growing potato plants is a boon. On the other hand, the company that owns the patent on the New Leaf requires certain commitments from the farmer, including the agreement not to pass the potato seeds on to other growers. This ultimately ties the farmer to the seed company in a way that has never been true before.

The Bt toxin engineered into the New Leaf is also a common pesticide used by organic farmers. They spray Bt onto their plants, where it is eventually washed off and decomposes quickly in the environment. Imitation is considered one of the highest forms of flattery, but the biotech industry's embracing their decades-old strategy does not please many organic farmers, for an important reason.

Pesticides have the same effect on insects that antibiotics have on microbes—they select for resistance. And resistant pests would make the organic farmers' practices useless. Researchers have already spotted a few beetles resistant to the Bt toxin. The massive quantities of Bt placed in the fields via the New Leaf and similarly engineered plants will no doubt provide an environment guaranteed to hasten the selective process for resistant strains. The organic growers and others surmise that Bt will not survive long as a "magic bullet."

The consumers who might purchase the genetically engineered New Leaf have their reservations as well. Although the toxin has no known effect in humans, it has never been eaten on such a large scale. Since the toxin is genetically engineered into the plant, it appears throughout all the plant tissue, including the potato itself. Many people have reservations about introducing the toxin into the food web in such an untested and massive manner. Such reservations underlie the refusal of some countries like Germany and France to import genetically engineered foods.

The New Leaf and the popular reaction to it expose some of the issues surrounding genetically manipulated crops. Employing microbial tools to genetically manipulate plants has rapidly become commonplace in many of the major agricultural businesses throughout the world. The year 2000 will mark only the fifth year that genetically engineered plants have been on the market, yet some 45 million acres of American farmland have already been planted with genetically altered crops.

"Super beetle" (certified pesticide resistant)

The monarch butterfly's breeding range includes the corn growing regions of North America—regions where Bt corn has become an important agricultural crop.

The Politics of Plants

It's likely the public debate surrounding the potential consequences of our ability to manipulate the genes of plants will continue well into our future. The creation and use of genetically engineered plants (and microbes and animals, for that matter) are issues that raise considerably divergent viewpoints among those who will be directly affected.

Consider Bt corn and the plight of the monarch butterfly. Since all the cells of the corn plant contain the gene for the insecticide, all the cells are capable of producing the toxin. This includes the corn's pollen cells, which have evolved to disseminate the plant's progeny to other fields. All the cells of the corn plant contain the gene for the insecticide, including the corn plant's pollen. Since the pollen might alight on other plants nearby, researchers tested its killing effect by sprinkling it on milkweed and feeding it to the caterpillars of the beloved monarch butterfly. The caterpillars promptly died.

This finding engendered immediate concern among environmentalists and other interested parties, some of whom declared that the corn should be banned from use. Assessing the risk presented to the monarch by the corn pollen in the wild is a difficult issue. The monarch's breeding range includes the American corn belt, but they don't feed directly on the corn itself. Their preferred food is the milkweed. If the corn pollen lights on the milkweed, some monarch caterpillars could certainly consume it.

Milkweed is the favored food for hungry monarch caterpillars. Whether the caterpillars would be endangered by stray Bt corn pollen lighting on the plant remains a difficult question.

Whether the pollen would spread outside the corn field is unknown, and the flip side of the coin is that some insecticide will be used in any event. So is there more risk in confining the insecticide to the plant's cells, even though they may be blown outside the growing field, versus simply spraying it on? And how do we assess those risks in the face of so much uncertainty?

Victor Masona's virus-resistant cassava plant and the Bt toxin-containing corn are but two stories among many. They offer us an important insight about our coming years on this planet. Many people believe that our future demands for food cannot be met by simply continuing what we are doing. The benefits of the Green Revolution will continue to accrue, but the rate at which we are increasing our output of food is falling behind the rate at which our population is growing. Genetic technology, born out of our microbial alliances, is providing a way to address some of this shortfall, and Zimbabwe, along with the rest of the world, must decide whether to embrace such solutions.

Beyond Food

The ability to introduce genes into plants has opened a door to what may ultimately be an even more significant advance. What if you could take a crop like tobacco and turn it into a green manufacturing facility for useful products? Or what if you could provide a vaccine to the world's children by simply having them eat their vegetables?

New genes introduced into tobacco plants may turn a health hazard into a factory of the future.

Both scenarios are possible and even probable. Scientists have already successfully created tobacco and other plants that produce everything from human proteins like interferon and antibodies to granules of polyester used to manufacture plastics. The engineered plants can produce very large amounts of such materials at very low cost.

Edible vaccines that can immunize against diseases like cholera and typhoid fever are under development and offer a tremendous opportunity to improve world health. A potato or banana producing the appropriate antigens to immunize against the major agents of diarrheal infection could help control one of the leading causes of infant mortality. Needing no refrigeration or special equipment, the vegetables could be delivered to the most remote places on the planet. Such engineered plants will very likely be available for use within the next five years. And microbial partners will play a large role in the actual engineering.

The Dark Side

Not all partnerships with the microbial world hold promise for humanity. The uses of microbes as agents of war and more recently for bioterrorism stand out starkly amid the benefits we can derive from the microbial world.

The notion that one could use microbes as weapons of destruction is not a recent idea. The first recorded use of biologic agents goes back to the Roman soldiers' practice of using dead animals to foul the enemy's water supply. The illnesses resulting from drinking the water were a consequence of infectious microbes from the dead animals contaminating the water.

Bacillus anthracis

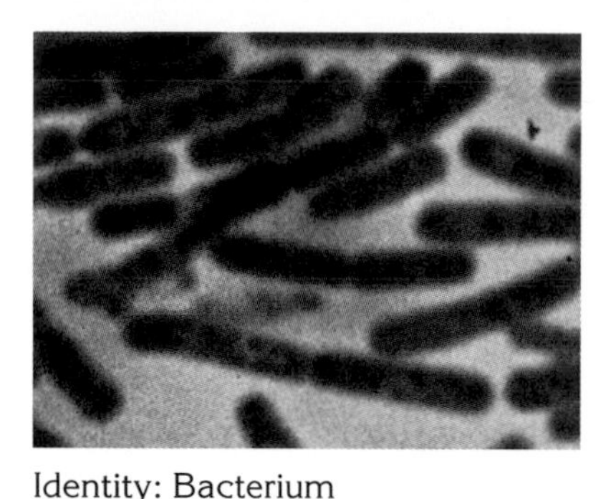

Identity: Bacterium
Residence: Soil contaminated by infected cows, sheep, pigs, goats, or horses
Favorite pastime: Going dormant in the dirt
Activities: A member of The Lethal Agents, *Bacillus anthracis* causes serious disease to people and other animals who may encounter the bacterium; a popular choice for biologic warfare and terrorism.

Microbes became a more specific weapon in humanity's arsenal soon after they were identified as the agents of infectious diseases. As early as 1763, the British supplied blankets known to be contaminated with smallpox virus to the Native Americans of an Ohio tribe in an attempt to quell rebellion. Other examples abound.

The world wars of the 20th century introduced the threat on a global scale. Russia, Japan, and the United States, along with several other countries, began to invest heavily in both offensive and defensive strategies that employed biological weapons. The agents under study were especially suited to the demands of military use: they were easily disseminated and lethal or highly toxic. Laboratory tests on animals indicate that 10 grams of anthrax spores, a bacterium causing a rapidly fatal lung infection, can produce as many casualties as a metric ton (1 million grams) of nerve agent.

Although the use of biological weapons has been condemned, and international treaties have been put in place to ban their use and to punish those who use them, biological warfare remains a concern today. Production of

biologic agents such as anthrax, smallpox, and botulinum toxin is simple and inexpensive. The biotechnology revolution provides ways in which specialized microbial agents can be readily created.

The United States unilaterally renounced biologic weapons in 1969, closing down its laboratories at Ft. Detrick, Maryland, and destroying all its stocks. Many countries have ratified the 1972 Bacteriological and Toxic Weapons convention. Today, no countries admit to having currently active, offensive biologic weapons programs. Yet recent history indicates that Iraq had an active program in place during the Gulf War, and Russia may have also maintained an active program. Another 10 or 12 nations are suspected of either having a program or having intentions to build one.

Military use aside, there remains the threat of bioterrorism. Biologic weapons are often referred to as the "poor man's nuclear bomb." They are easy and inexpensive to make. Thus far, fortunately, their use has been extremely limited. There is only a single act of bioterrorism on record in the United States. In 1984 in Dalles, Oregon, the members of the local Rajneesh cult intentionally contaminated salad bars with *Salmonella.* The act resulted in 751 people becoming ill, although no one died.

For avid supporters of biodiversity, there may be a silver lining in the dark clouds surrounding all the recent attention on biologic weapons. Russia and the U.S. hold the only known remaining stocks of smallpox virus, the agent of a deadly disease declared eradicated by the World Health Organization in 1980. The two countries had scheduled to destroy the stocks by July 1999—an act that would have been the first intentional extinction of a species. Under pressure from the President of the United States, the members of the World Health Assembly agreed to a postponement until 2002. The primary reason given by the President was national concern about the existence of undeclared stocks of the virus—stocks that would have been retained for only one purpose—biologic warfare.

Variola

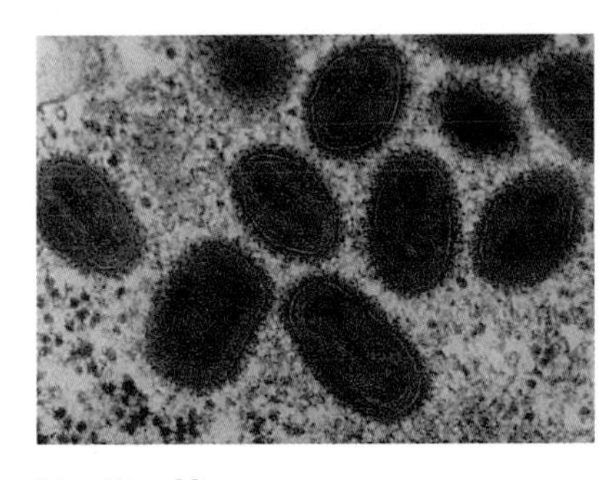

Identity: Virus
Residence: Officially restricted to a frozen state in two laboratories, one in the United States, the other in Russia
Favorite pastime: None, any more
Activities: A member of The Lethal Agents, variola causes smallpox; this virus is the first to be eradicated as an agent of human disease, but fears remain that secret stocks may be used for deadly purposes.

Of Microbes and Humans

What lies ahead for the human species on this planet we call earth? Despite centuries of scientific progress, we still face monumental challenges. Infectious diseases still take the lives of more than 50,000 people a day worldwide, and emerging antibiotic resistance compromises some of our most valuable treatment advances. The environmental legacies of our past practices, and the legacies we are creating today, expose millions of people annually to potentially toxic chemicals and nuclear waste. At the close of the 20th century, there are still 790 million people who are food poor.

One thing is clear. Our exponentially increasing knowledge and understanding of the microbial world offer us opportunities to improve the human condition in the coming decades. That knowledge and understanding may also present us with new challenges.

Microbes will provide us with new antimicrobial compounds for treating infectious diseases, including some infections caused by pathogens resistant to our present antibiotic armamentarium. Still, antibiotic resistance, brought on in part by our own actions, will continue to emerge, tempering our success. We need to rethink our widespread use of antibiotics for non-disease purposes or find new approaches to prevent a worsening of the threat from antibiotic-resistant microbes.

Microbes offer us new approaches to cleaning up contaminants that we have left behind, changing some toxic chemicals and other pollutants into benign substances and helping to restore our damaged environment. Yet, the worst environmental legacies of our past manufacturing practices and nuclear industry remain beyond the reach of even the most voracious microbes. In the worst cases, we must simply store our toxic wastes and await the discovery or creation of a solution. The chances are good that the solutions will lie in the microbial world.

Microbes, harnessed as miniature factories, can provide us with cleaner and more efficient ways to produce everything from antibodies to plastics and fuel, perhaps helping us avoid repeating the environmental sins of our past. Yet, we will need to find a way to ensure that the economic incentives offered by these microbial processes are made clear and are sufficient to encourage both developed and developing countries to employ them.

Microbes can equip us with tools to improve the nutritional value of the food we grow as well as the way we grow it, ushering in the next Green Revolution. Yet, feeding the citizens of our global village will continue to challenge us if the human population continues to expand, and we will need to understand and plan for potential unintended consequences from the mass use of genetically engineered stocks.

In the end, earth's resources *are* finite. Although microbial technologies will allow us to prevent deaths from infectious diseases, increase our production efficiency while sparing our environment, and improve our food supply, they may also mask the finite nature—the scarcity—of earth's resources. Technology does not replace human responsibility, and at some point, we must all find a way to balance our demands against those of the rest of earth's inhabitants.

We are part of the complex web of life on our planet—not above it, and not beyond it. Our choices will influence that web in ways we are only beginning to understand. Our actions will change the future. Let us hope that the lessons we have learned from and the partnerships we have formed with earth's oldest and most successful inhabitants will help us attain the knowledge to make the right choices.

Small Things Considered

Most microbial matters chronicled in this book are examined at the level of the whole organism or of communities of organisms out there contending with their environment. The various commentaries in this section look *inside*—at the molecules and patterns of interaction of molecules that underlie the behavior of these remarkable cells: their ability to "feel" their environment, to communicate, to get and use food to grow and divide. It is at this, the molecular level that the unity of biology becomes most apparent. These microbes, and all the plants and animals derived from them over the vast span of evolutionary time, do things pretty much the same way. Indeed it is here, among the molecules, that we most clearly confront our heritage.

Information

Information, for a living cell, is of two distinct kinds:

1) A cell reads, or perhaps more accurately "feels," the world around it; that is, it takes in information and responds to it. (This domain of information resides in the cell's proteins; see The Amazing Behavior of Proteins, below.)
2) A cell maintains a bank of historical memory, in the form of coded instructions. This is its genetic information.

Genetic information depends on difference to convey the many instructions needed to create a living cell.

A chain of unvarying units contains no information.

A chain of two or more different units can convey an infinite amount of information, given sufficient length.

— · · — · — — · · · —	Morse code (2 units)
0100110 01010111011	Computer language (2 units)
To be or not to be	English (26 units)

Life's genetic information is DNA (deoxyribonucleic acid)—a chain of chemical units, called nucleotides, abbreviated A [adenylic acid], T [thymidylic acid], G [guanylic acid], and C [cytidylic acid] (as depicted on p. **70**).

ACGTATGGCAATT	DNA

The sequence of units in Morse code, computer language, and English evolved over recent times and is interpreted by human minds. The sequence of nucleotides in a DNA chain evolved over vast spans of time and is interpreted by machinery in cells to produce the proteins of a living organism.

It takes a considerable amount of information to make even the simplest of cells. The DNA of the tiny bacterium *Escherichia coli* is composed of two long, intertwined chains, each made up of about 4.7 million A's, T's, C's, and G's. This is equal to the number of letters in eight average-length novels. The double-helical molecule is a compact, continuous circle, but if it were opened and stretched out, it would be 500 times longer than the bacterium.

Operationally, one of the DNA chains is divided into genes, each averaging some 1,000 nucleotide units in length—the equivalent in letters of a long paragraph. The 4.7 million units in *E. coli*'s DNA chain accommodate in excess of 4,000 genes.

DNA's doubleness is a key to its immortality—its copy-ability, if you will. This doubleness depends on the fact that every A in one chain is loosely bonded to every T in the opposite chain and, similarly, every G to every C. This complementary pairing of the chains allows them to separate, have new units laid down along each strand, and so be perfectly reproduced. Once the double DNA chains are so duplicated, the cell can divide, forming two daughter cells, each with its own double DNA chain. Thus, information passes from generation to generation.

The Protein-Making Machinery

DNA does not envision its product—the living being. It is not a blueprint or an image. It is, rather, a recipe—a set of instructions for making protein molecules that will provide form and function to the living being.

The machinery of protein synthesis reads the instructions—the sequence of nucleotides in genes—and translates them into an amino acid chain—a protein (see p. **52**). The process is analogous in principle to a machine translating a stretch of the two units of computer language into a sequence of the 26 letters of English. In the case of the protein-making machinery, one or more triplets of nucleotides represent each of the 20 amino acids to be linked into protein chains. Thus, what we call the genetic code is written in three-letter words.

Here are the key steps in the process:

1) Each of the 20 amino acids to be linked together to make a protein is first given a chemical signature, or identity, by being linked up to a special RNA molecule called *transfer RNA*.
2) One strand of a gene's DNA is copied into an RNA replica of itself, that is, a single strand of nucleotides called messenger RNA. *Messenger RNA* is an expendable gene copy.
3) The messenger RNA is "read" triplet by triplet—"word" by "word"—by a complex machine called a *ribosome* (p. **61**). Transfer RNAs carrying their amino acids enter the messenger RNA production line on the ribosome. A triplet of nucleotides—the anticodon or "anti-word"—on each transfer RNA matches up with a complementary triplet of nucleotides—the codon or "word"—on the messenger RNA.
4) Once the amino acid is located correctly in the sequence dictated by the messenger RNA, it is chemically bonded to its previously deposited neighbor, and the ferrying transfer RNA departs.

Thus, a protein is built from one end to the other, one amino acid at a time. Upon reaching the end of the message, the finished protein drops off and folds up into its final functional structure. There are many thousands of ribosomes in a single cell carrying out this process simultaneously, producing hundreds of different proteins.

The Amazing Behavior of Proteins

Proteins mediate the essential functions of life—movement, the production and use of energy from food, self-construction, self-regulation, and replication. Proteins make up most of a cell's infrastructure and are its workers, its machinery.

Remove all the water from a cell, and 80% of what's left is protein molecules—thousands of different kinds of them. A living creature is the embodiment of the coordinated activity of these molecules. Differences between proteins are what make the differences among us and between us and other life forms. Yet protein-guided processes show remarkable similarities across the whole spectrum of life.

How do proteins do it? The essence of their wizardry is chemical recognition; that is, the ability of each protein to interact selectively and specifically with only one or a very few other molecules. A binding site—like a cavity or pocket—on the protein's surface recognizes a particular molecule, as a lock recognizes a key, and holds it with weak, close-range chemical bonds. This recognition event can be put to many clever uses. The crucial ones are described below.

Structural Components

Much of a cell's infrastructure is protein. Proteins are basic components of the membranes that surround a cell and hold its contents together. Proteins make up the devices in the cell membrane that act as channels through which nutrients enter the cell and wastes leave. Proteins are the basic building blocks of the ribosomes, the cell's protein-manufacturing machinery. Contractile proteins, the progenitors of muscles in higher organisms, mediate cell movement. Proteins make up flagella, long whiplike appendages that are the microbial equivalent of propellers powered by a rotary engine, also made of protein. And when bacterial cells physically contact each other for purposes of lateral gene transfer, they do it with a connecting tube (a pilus) through which DNA passes. That, too, is made of protein.

Workers

Certain proteins called *enzymes* are catalysts, or facilitators of chemical reactions. They are the cell's workers; they put things together or break them apart. Each enzyme has at least one binding site. Like a lock and key, an enzyme's binding site recognizes and locks in another molecule (called a substrate), causing a chemical modification of that molecule, either breaking it into smaller fragments or attaching it to something else. The altered molecule(s) is released, and a fresh one enters and is similarly processed in a rapidly repeating cycle. For example, an enzyme can grasp a sugar molecule, break it into smaller parts, and in the process extract energy from it, all in a fraction of a second. Assisted by that energy, other enzymes can facilitate the linking of small molecules to each other to construct larger molecules.

Regulators and Orchestrators

Critical to cellular processes is the capacity not only to catalyze large numbers of chemical reactions but also to manage them, i.e., control the rates at which they operate. Many enzymes have a signal-sensitive site on their surface which, when it binds a signal (usually a small molecule)—again in a lock-and-key type fit—responds by speeding up or slowing down its work.

Proteins also act as "genetic switches." By grabbing hold or letting go of specific parts of the DNA, they activate (turn on) or deactivate (turn off) specific genes, thereby controlling whether messenger RNA is made and hence whether the gene-encoded protein is produced. This switching enables the cell to selectively orchestrate its responses to the world around it. For instance, a bacterium can turn on a gene that governs the synthesis of an enzyme that allows it to use a food that suddenly appears in its neighborhood; conversely, it can shut off the gene if the food is not available. Clearly, economy of operation is critical to survival.

Alteration of the function of these so-called regulatory proteins can have widely distributed effects on the performance of many proteins, dramatically affecting the size, shape, or internal function of an individual (as illustrated on p. **53**). To use a dramatic example from our world of multicellular organisms, 98% of the proteins in humans and chimpanzees are identical. The root of the difference lies with the few regulatory proteins that control the development of systems that both creatures have in common—musculature, bone, brain, etc. While variation in microbes is not as

readily apparent as that between human and chimps, the same principles apply: differences in proteins, particularly in proteins with a regulatory role, determine differences in form and function.

Communicators

Another fundamental role of proteins and their unique recognition capacity is information—processing among cells and between cells and their environment. Cells receive information in the form of chemical signals. These are detected at the cell's surface by specific proteins called *receptors*. (See, for instance, p. **92**). Receptors are embedded in the cell's outer membrane. One end of the protein chain sticks outside the cell, ready to receive chemical information in the form of a signal molecule. The other end projects into the cell's interior, ready to influence internal events. The outsides of cells bristle with many different outward-facing receptor ends, each of which recognizes only one kind of signal molecule.

When a signal contacts a receptor end, the receptor protein changes shape all along its length, from the outside of the cell through to the inside. The change in the inward end initiates a chemical event: an internal response to the signal.

An example: when a motile bacterium senses food—say, sugar—in its environment, it means that its special food-recognizing receptors are binding sugar molecules. The consequent shape change in the receptor communicates to the cell's interior, setting up a series of chemical events that starts a tiny rotary motor. The motor activates a whiplike appendage, a flagellum, on the outside surface of the bacterium, which propels it toward more sugar molecules.

Distinguishing Self from Non-Self

If any two cells in your body had a chance encounter, they would recognize each other. Like a secret handshake, a molecular "fit" on the surface instantly informs each cell that it has bumped into one of its own kind.

This skill at distinguishing self has enormous implications for our health. It's how our cells know when intruders are present.

Certain proteins (H proteins) sticking out on the surfaces of the cells of each individual are different (except in identical twins). (The blue molecules on the surface on the macrophages on p. **121** are H proteins). They are what identify each person's cells as unique. Lymphocytes called T cells are a part of the body's immune system. They mature in the thymus gland and then are released to circulate around the body, searching for foreign invaders. While in the thymus before their release, they come constantly into contact with the H proteins of the thymus's own cells. Protein receptors on the surface of these T cells that fit comfortably with the thymus's H proteins survive and mature. The rest self-destruct (a phenomenon called apoptosis). Thus, the mature T cells now dispatched to search for trespassers will recognize the familiar H protein on the surfaces of the body's many other cells as self and pass on.

But H proteins have another key function. They stick to foreign proteins presented on the surfaces of cells. This H protein–foreign antigen combination is what the T cell recognizes as *enemy*. And it is this H protein–foreign antigen–T cell receptor complex that triggers the T cell's immune response to the enemy.

This same mechanism underlies the rejection of the transplanted tissue from another individual. The H proteins on the surfaces of transplanted cell surfaces are seen by T cells as non-self—just as they see the self H protein–foreign antigen complex as non-self—and they immediately set in motion the processes of immune defense.

This is the answer to what had long been a great mystery of biology: why would animal immune systems have evolved the means of rejecting the cells of other individuals when those systems had never had any contact with those cells?

Getting and Using Energy

Energy—the ability to do work—is the enabler of life. Life on our planet receives most of its energy from nuclear fusion in the sun, with smaller contributions from the earth's hot core, from energy-rich inorganic materials in the earth. Incoming energy passes through earth's food chain before departing as heat.

Most of the energy for life is captured through photosynthesis, which provides the energy to transform CO_2 and H_2O into sugar, life's principal energy reservoir. Sugar is a hydrocarbon, i.e., a chain of carbon atoms linked to each other with hydrogen atoms attached.

Carbon and hydrogen atoms are available in large quantities on earth. They exist primarily in their most stable, least reactive, lowest energy states: carbon dioxide (CO_2) and water (H_2O). The process of elevating carbon and hydrogen from their low-energy bondage in CO_2 and H_2O to their energy-rich state in sugar requires the input of a lot of energy.

How does life do it? Electrons are the key. These tiny, negatively charged particles orbiting the nucleus of every atom can be mobilized to create the biological equivalent of an electric current. The hydrogen atom (H), in particular, plays a big role. Made of a single electron and a single proton, it can readily donate its electron to other molecules and then retrieve it to become H again.

Here's the sequence of events in the making of sugar by photosynthesis. Photons of light hitting pigment molecules (light attractors like chlorophyll) bounce electrons out of their customary orbits, producing a flow of electrons along a chain of electron-accepting molecules. (The electrons displaced from the pigment molecules are replaced from the hydrogens of water.) This flow, or current, is harnessed by a special protein in the membrane to make the energy-rich molecule ATP (adenosine triphosphate). Thus, light energy has been transformed into chemical bond energy.

The electrons, having been instrumental in the production of ATP, recombine with protons, and end up as hydrogen atoms again. These hydrogen atoms are attached to special carrier molecules which put them in a uniquely reactive state. They enter a chemical dance with CO_2, aided by enzymes and energized by the ATP they've helped to generate, and become sugar. The sugar reservoir is now available both to the organisms that created it and to the freeloaders who eat the creators.

There are two processes by which all organisms—both sugar creators and freeloaders—extract energy from sugar to make the ATP they need to do all the work of living: *fermentation* and *respiration*.

Fermentation, in which enzymes wrestle small amounts of ATP from sugar molecules by breaking the molecules in half, is life's most ancient way of generating energy. Respiration evolved well over 2 billion years ago, when oxygen—the waste product of photosynthesis—began to accumulate in the earth's atmosphere. Respiration repetitively extracts electrons from pieces of sugar produced by fermentation, and the resulting flow produces ATP. The electrons, having done their job, are combined with oxygen to make H_2O, and the carbons of sugar are combined with oxygen to make CO_2.

Overall, life, using sunlight, elevates CO_2 and H_2O to high-energy sugar and oxygen, consumes sugar to live, and gives back CO_2 and H_2O. Energy, then, flows downhill while materials constantly recycle.

Mutation

A mutation is a permanent, heritable change in DNA. It is most often the substitution of one nucleotide for another, but it may involve the addition or removal of one or more nucleotides, the movement of stretches of nucleotides from one location to another, or the insertion of a segment of foreign DNA into an organism's genome. These events occur spontaneously and randomly at any location in the genome. Mutations are expressed when genes are translated into proteins as alterations in the protein's amino acid sequence (as shown on p. **52**). Therefore, the function of a protein synthesized from the affected genes may be altered. (Mutations may also change an amino acid in a protein *without* changing its function.)

In microbes, all mutations are duplicated when their DNA is duplicated and so are transmitted to the daughter cells when the cell divides, and so on, generation after generation. (In sexually reproducing organisms, mutations that occur in the DNA of the cells that make sperm and eggs are passed to the next generation; mutations occurring in body cells are passed only to that cell's daughters but *not* to the next generation.)

There are many causes of mutation. Most are the result of errors made by the enzymes that copy DNA—like typographical errors. Other causes are direct chemical damage to DNA produced by the ever-present cosmic radiation, by ultraviolet light, or by X-rays and chemicals in the environment. Mutation-producing agents are called mutagens. All organisms from microbes to the largest multicellular creatures are subject to their effects. Mutations arise constantly as an integral part of being alive.

In considering mutation and selection as the instruments of evolution, we can't help but marvel at the stability of the genomes of living creatures over time. Indeed, the rate at which stable, permanent mutations accumulate in DNA is estimated to be roughly only one nucleotide change in 1 billion nucleotides every 200,000 years! One way this has been measured is to count the number of nucleotide differences in the same gene taken from two different species and relate this to the number of years that have elapsed since the two diverged from a common ancestor. The time elapsed can be deduced from fossil records.

Since by far the majority of mutations are likely to produce a damaged protein (or RNA), and thus lead to an organism's, and its offspring's, failure to survive, we are measuring only the few mutations that have led to success through natural selection. Only the occasional organism's DNA is altered by a mutation that improves its function. But the faster an organism's generation time, and the larger its population, the greater is the potential for change over time.

A single bacterium with some 4,000 genes would be expected to suffer a single stable nucleotide change somewhere among those genes every 250 generations. A population of as few as 250 bacteria, then, would experience a mutation once every time the population doubled. It should thus come as no surprise that, over their 3-plus billion years of existence, microbes have evolved most of the devices for living which we multicellular creatures, evolving much more slowly over a much shorter span of evolutionary time, have co-opted for our own purposes.

From Horizontal Gene Transfer to Internal Gene Tinkering

The rapid generation time of microbes and their ability to pass around their genetic information give them an advantage when it comes to innovation (see p. **83**). These two factors have made it possible for microbes to explore almost every imaginable avenue for exploiting their environment. Multicellular organisms began to emerge late in evolution, only 600 million years ago (15% of evolution's total span). They differed in many ways from their single-celled ancestors, but importantly, they were no longer able to exchange genetic information with quite such abandon. Coupled with the much longer time necessary to produce a new generation, multicellular creatures were at something of an evolutionary disadvantage. They needed to devise new ways to survive and thrive.

Multicellular organisms seem to have accomplished this primarily by tinkering with the genes they already had. First, they began duplicating some of their genes, or parts of them. The extra copies were then available to be modified into genes that coded for proteins with related, but different, functions. Genes became separated into segments: exons that coded for parts of proteins and introns that could be discarded. The exons were combined to make mosaic genes that coded for proteins with new duties. The key part of each new gene was the sequence of nucleotides that prescribed a protein's *recognition sites:* that is, the part of an enzyme that recognizes its substrate, the part of an antibody that recognizes an antigen, the part of a regulatory protein that recognizes a particular site on DNA, the part of a receptor that recognizes a signal. Other parts of these genes need not have been as precisely specified.

The combinatorial potential of such shuffling of gene segments produced an enormous variety of useful proteins.

As we learn more and more about their structure, most proteins in multicellular organisms today appear to be descended from a limited number of ancestral types. We suspect that the burst of all kinds of new multicelled life forms some 600 million years ago was, in part, the result of these new modes of generating genetic novelty.

How to Make Billions of Antibodies with a Little DNA

We saw on page 122 how antibodies attach to, and sometimes disable, invaders. Or at least they identify the intruder until help arrives. To be prepared for the vast variety of potential intruders, our immune system must produce an equally vast variety of differently shaped antibodies. This is a remarkable ability.

An antibody is a protein consisting of two long and two short chains linked together to make a single large Y-shaped molecule (see p. **122**). The DNA of humans and vertebrates contains several hundred genes that code for these antibody chains.

In general, the DNA of every cell in the same organism is identical. But B cells, the makers of antibodies, break the rule: the genes that code for the short chains are very different from cell to cell. This is because, as B cells mature, the genes coding for the short chains get randomly cut up, shuffled, and recombined.

The structure of the short chain of the final antibody made by each B cell is, therefore, a matter of chance. Each individual B cell's recombinational events produce only one antibody variant. But, since billions of B cells can be produced, each with its own unique antibody, the potential for matching any possible antigen entering the body is enormous. So, a small amount of DNA, by reshuffling segments of itself, can produce an enormous number of antibodies.

The antibody a B cell makes is exhibited on its surface. Once it recognizes and binds to a particular, precisely fitting foreign antigen, it signals the protein-making machinery within the B cell to make large quantities of that antibody.

Cloning

Cloning is copying: producing many identical individuals or molecules from one or a few. A bacterium clones itself when it multiplies—one becoming two, two becoming four, etc.—until there's a mass of many millions of identical bacteria: a clone.

A clone of DNA is many copies of any length of DNA made by repeatedly duplicating it. Bacteria, as they clone themselves, also clone their DNA. Furthermore, if a piece of foreign DNA, such as a gene or genes from a human, is inserted into a bacterium, it will be copied along with the bacterium's own DNA as it divides into two daughter cells. The resulting cloned foreign DNA may then be recovered from the clone of many millions of bacterial cells.

This method of producing multiple copies of DNA by introducing novel genes into living bacteria and using the bacteria as copying machines is called *recombinant DNA technology* (see example on p. **147**). Recombinant DNA was the key to opening up the science of genetic engineering in the 1970s. Cloning made it possible to obtain any particular DNA in large enough quantities for detailed chemical analysis and for genetic experimentation.

It is now possible to clone DNA without the assistance of live bacteria. The bacterial enzyme responsible for replicating DNA, called DNA polymerase, can be isolated from bacteria, purified, and set to work on DNA in the test tube. It will copy any DNA when supplied with plenty of DNA's four nucleotide-building units—A, T, G and C. The process is called polymerase chain reaction or PCR. (See p. **70.**)

The cloning of a multicellular organism such as a sheep is, in principle, an extension of the principle of cloning of a single-cell bacterium. Every cell of an animal (with the exception of a few types that naturally shed their DNA as they mature, such as red blood cells) contains a full complement of DNA—all the information to recreate that individual. Theoretically, then, *any* cell, if properly induced and cultivated, can multiply and differentiate to become a complete new individual—in this context, a clone.

The main impediment is finding the means to free the DNA of the starting cell from the restraints imposed upon it during its earlier differentiation when it became part of the individual from which it was taken. A skin cell, for example, while endowed with all the information to become a complete individual, has undergone changes that suppressed the information within it to become a brain or liver cell. This hurdle has been leaped in a growing number of animals.

Getting Started

The earliest events in the emergence of life are likely to have been the assembly of potential information-carrying molecules and the machinery to replicate them. Both, remarkably, can be achieved by simply making chains. Today's information chains are DNA; its machinery chains are RNA and protein. Back then, almost 4 billion years ago, all three functions may have been performed by a single kind of molecule: RNA.

We suspect this because RNA is similar enough to DNA in its capacity to act as a perfectly good repository of information. And we now have a growing body of experimental evidence that RNA can fold into intricate shapes, the way proteins do, and act like a primitive biological catalyst, i.e., an enzyme. Thus, RNA could have acted both as an information carrier *and* as a protein.

On early earth—a hot, soup-like environment filled with a variety of chemicals available for use as building materials and sources of energy—RNA chains could have emerged and grown. Floating about as randomly assembled pre-information, some may have been replicated with the aid of others acting as primitive catalysts.

With a near-infinite supply of chain-building units and vast stretches of time, the inevitable growth of chains plus copying mistakes would produce chains in enormous varieties. Those that were more efficient at copying and being copied would thrive, produce more of themselves, and inevitably interact with other components to produce even greater efficiency, variety, and complexity.

Eventually, by means our imagination is stretched to the limit to envision, these processes of natural selection got enveloped into tiny private worlds of their own, surrounded by a membrane: a major step toward becoming a cell. This primeval cell would serve to keep the molecules participating in the self-replication process in proximity to each other; it would devise the means to draw building blocks and sources of energy into itself from outside; and it would need to replicate itself in synchrony with the replicating molecules within.

At some point, DNA and protein would relieve RNA of some of its burdens. The vehicle that started this prodigious journey to the present would by this time deserve a name: bacterium.

Index

F

G

H

I

J

K

L

M

N

O

P

Photographic credits

i: earth from space, Jet Propulsion Laboratory. ii: viruses, Photo Researchers; rotavirus, Courtesy of Wah Chiu and B.V.V. Pasad. xiv: Jet Propulsion Laboratory; volvox, Bruce J. Russell, BioMedia Associates; microbes, Courtesy Carl Woese; bacilli, ©Oxford Scientific Films. Page 2: barge, ©Visuals Unlimited, Bernd Wihich. Page 3: *Methanothrix,* J. P. Touzel et al. Page 4: planet, ©Visuals Unlimited, Sci Vu. Page 6: *Chlorella,* J. L. Van Etten et al.; microscopic kelp, ©Visuals Unlimited, CABISCO; kelp, ©Film Bank. Page 7: methanogen, S. Fukuzaki et al.; termites, ©Visuals Unlimited, Bill Beatty; termite gut flora, ©Visuals Unlimited, M. Abbey. Page 8: *Pedomicrobium,* R. Gebers and M. Beese. Page 9: Courtesy Kenneth Coale, Moss Landing Marine Laboratory. Page 12: *Rhizobium,* Courtesy Frank Dazzo. Page 13: root nodules, U.S. Department of Agriculture. Page 16: *Aspergillus,* B. Jahn et al. Page 17: mouse sequence, Cinenet/©1999 Ken Middleham. Page 24, leaf-cutter ants, ©Visuals Unlimited, George Loun. Page 25: caterpillar, ©Visuals Unlimited, William Paimer. Page 26: waves, ©Visuals Unlimited, Doug Sokell. Page 30: sea clam, ©Visuals Unlimited, Valorie Hodgson. Page 31: coral, ©Visuals Unlimited, William Ober; anemone, ©Visuals Unlimited, Marty Snyderman; sea fan, ©Visuals Unlimited, Hal Beral. Page 32: green sheet coral, Secret Sea Visions. Page 33: bleaching brain coral, Secret Sea Visions. Page 36: deep vent community, ©Woods Hole Oceanographic Institution. Page 38: tube worms, ©Visuals Unlimited, WHOI-D. Foster; deep vent community, National Geographic Television. Page 40: *Exxon Valdez,* seal, both Courtesy Alaska Resources Library and Information Services. Page 41: worker, ©Visuals Unlimited, Will Troyer. Page 44: Bruce J. Russell, BioMedia Associates; Quill Graphics. Page 47: trilobite, ©Visuals Unlimited, K. Lucas. Page 48: fossil microbe, Courtesy J. William Schopf; stromatolites, ©Visuals Unlimited, B. Rogers. Page 49: *Cyanobacterium,* Bruce J. Russell, BioMedia Associates. Page 65: smokers, ©Woods Hole Oceanographic Institution. Page 68: DNA, ©Visuals Unlimited, K. Murti. Page 69: microbial biofilm, L. R. Berger and J. A. Berger. Page 72: eukaryotic cell, BioMedia. Page 73: six Eukarya, video microscopy by Jeremy Pickett-Heaps, Cytographics. Page 76: mitochondrial origin, Quill Graphics. Page 78: *Synechococcus,* ©Visuals Unlimited, Elizabeth Gentt. Page 80: mushrooms, Cinenet/©1999 Ken Middleham; mushroom cells, ©Visuals Unlimited, Stan Flegler. Page 81: animal cells, Cytographics; penguins, ©Frans Lanting/Minden Pictures. Page 88: *L. pneumophila,* Photo Researchers; hantavirus, Photo Researchers; dividing microbes, Quill Graphics; Ebola virus, Photo Researchers. Page 93: *Propionibacterium (Corynebacterium),* Dennis Kunkel, University of Hawaii. Page 94: *Staphylococcus epidermidis,* W. Ziebuhr et al.; *S. epidermidis* on catheter, Olson et al., *J. Biomed. Materials* **22:**485-495, ©1988, Reprinted by permission of John Wiley & Sons, Inc. Page 95: *Porphyromonas,* L. Du et al.; *Lactobacillus,* K.-L. G. Ho et al. Page 97: *H. pylori,* ©Visuals Unlimited, V. Burmeister. Page 98: *E. coli,* ©Visuals Unlimited, K. G. Murti. Page 103: Group A *Streptococcus,* ©Visuals Unlimited, George Musil. Page 106: *Plasmodium,* I. W. Sherman (ed.), *Malaria: Parasite Biology, Pathogenesis and Protection,* ©1998 ASM Press, Washington, D.C.; African hospital, Courtesy Ann Nelson. Page 107: red blood cells, ©Visuals Unlimited, Stan Fiegler. Page 108: Ebola virus, *L. pneumophila,* both Photo Researchers. Page 109: CDC workers, Centers for Disease Control and Prevention. Page 110: rabies virus, ©Visuals Unlimited, K. G. Murti; bat, ©Visuals Unlimited, T. Gula; rickettsia, ©Visuals Unlimited, Sci Vu; tick, ©Visuals Unlimited, A. M. Siegelman; *B. abortus,* ©Visuals Unlimited; cow, ©Visuals Unlimited, J. McDonald; *C. psittaci,* ©Visuals Unlimited, D. M. Phillips; parrot, ©Visuals Unlimited, J. McDonald. Page 113: HIV, J. L. Levy, *HIV and the Pathogenesis of AIDS,* ©1998 ASM Press. Page 114: *Borrelia burgdorferi,* A. Szczepanski and J. L. Benach; water plant, Oregon Public Broadcasting. Page 116-7: globe, from "Extreme Weather Events: The Health and Economic Consequences of the 1997/98 El Niño and La Niña," Paul Epstein (ed.), Center for Health and the Global Environment, Boston, Mass., ©1999. Page 117: mosquito, ©Visuals Unlimited, Glenn Oliver. Page 125: polyomavirus, R. D. Smith et al. Page 128: *Shigella,* N. J. Snellings et al. Page 131: Navajo drawing, Courtesy Andy Natanabah. Page 132: mother and infant, Oregon Public Broadcasting. Page 133: headline, Courtesy The WPA Film Library. Page 136: rotifer, video microscopy by Jeremy Pickett-Heaps, Cytographics; microbes, Bruce J. Russell, BioMedia Associates. Page 138: A. Fleming, Courtesy The WPA Film Library; *Penicillium,* ©Visuals Unlimited, George Musil. Page 149: *Streptomyces,* S. Omura. Page 153: methanotroph/TCE degrader, C. D. Little et al.; grazing microbes, BioMedia. Page 158: Courtesy IBM. Page 159: strain T1, P. J. Evans et al. Page 161: *Exxon Valdez,* Courtesy Alaska Resources Library and Information Services. Page 163: *Saccharomyces,* C. Costigan et al. Page 165: white fly, ©Visuals Unlimited, Sci Vu. Page 166: cassava mosaic virus, Rothamstad. Page 167: *Agrobacterium tumefaciens,* SPL. Page 172: butterfly and milkweed, Photos by Lincoln P. Brower. Page 173: tobacco plant, ©Visuals Unlimited. Page 174: *Bacillus anthracis,* ©Visuals Unlimited, A. M. Siegelman. Page 175: variola, ©Visuals Unlimited, Hans Gelderblom. Page 177: Milky Way, NASA. Page 178: volvox, BioMedia; bacteria, Quill Graphics; bacteria, Oxford; rotavirus, Courtesy of Wah Chiu and B.V.V. Pasad. Remaining photos ©Intimate Strangers, Inc. (Directors of Photography: Stuart Asbjornsen, Blake McHugh, Mitch Wilson)